Springer Series in Solid-State Sciences

Volume 191

The Springer Series in Solid-State Sciences consists of fundamental scientific books prepared by leading researchers in the field. They strive to communicate, in a systematic and comprehensive way, the basic principles as well as new developments in theoretical and experimental solid-state physics.

More information about this series at http://www.springer.com/series/682

Roberta Citro · Ferdinando Mancini
Editors

Out-of-Equilibrium Physics of Correlated Electron Systems

Springer

Editors
Roberta Citro
Dipartimento di Fisica
University of Salerno
Fisciano, Salerno, Italy

Ferdinando Mancini
International Institute for Advanced Scientific Studies "Eduardo R. Caianiello"
Vietri sul Mare, Salerno, Italy

ISSN 0171-1873 ISSN 2197-4179 (electronic)
Springer Series in Solid-State Sciences
ISBN 978-3-030-06944-5 ISBN 978-3-319-94956-7 (eBook)
https://doi.org/10.1007/978-3-319-94956-7

Printed on acid-free paper

This Springer imprint is published by the registered company Springer Nature Switzerland AG
The registered company address is: Gewerbestrasse 11, 6330 Cham, Switzerland

Preface

This volume "Out-of-Equilibrium Physics of Correlated Electron Systems" covers different theoretical aspects of the physics of non-equilibrium, ranging from the analytical to numerical and computational ones. It starts with the basic theory of master equations versus Keldysh Green's function formalism for correlated quantum systems out-of-equilibrium, passing on to the Gutzwiller variational method and non-equilibrium mean-field theory for superconductors, concluding with the dynamical mean-field theory of correlated electron models. The book offers a comprehensive overview of the recent advancements in the theoretical investigations of strongly correlated systems out-of-equilibrium with specific examples.

The book contains the notes of the lectures delivered at the "Twentieth Training Course in the Physics of Strongly Correlated Systems" held in Vietri sul Mare (Salerno, Italy) in October 2016. The course consisted of four lectures every morning, held by Profs. Enrico Arrigoni, Massimo Capone, Martin Eckstein, Stefan Kaiser, and afternoon activities (seminars delivered by the junior researchers, solving of specific problems, roundtable on hot topics, hands-on training on relevant numerical issues) aimed principally at promoting discussions between the attendees and the lecturers. The outcome of this type of course was a significant interchange of ideas among the participants thanks to both the enlightening morning lectures and the long afternoon sessions devoted to discussions.

The book has both introductory and pedagogical aspects that could be very useful for researchers entering the field of non-equilibrium physics of correlated electron systems and the readers will strongly benefit from the different overviews on the topic.

We wish to acknowledge the support of those institutions that made the course possible. The main sponsors of the event were the Department of Physics "E.R. Caianiello"—University of Salerno and the International Institute for Advanced Scientific Studies "E.R. Caianiello."

Fisciano, Italy — Roberta Citro
Vietri sul Mare, Italy — Ferdinando Mancini

Contents

Contributors

Enrico Arrigoni Graz University of Technology, Graz, Austria

Massimo Capone Scuola Internazionale Superiore di Studi Avanzati (SISSA) and CNR-IOM Democritos National Simulation Center, Trieste, Italy

Roberta Citro IIASS, International Institute for Advanced Scientific Studies "E.R. Caianiello", Vietri sul Mare, SA, Italy; Dipartimento di Fisica "E.R. Caianiello", Università degli Studi di Salerno, Fisciano, SA, Italy

Antonius Dorda Graz University of Technology, Graz, Austria

Martin Eckstein Friedrich-Alexander Universität Erlangen-Nürnberg, Erlangen, Germany

Carla Lupo Physics King's College, London, UK

Ferdinando Mancini IIASS, International Institute for Advanced Scientific Studies "E.R. Caianiello", Vietri sul Mare, SA, Italy; Dipartimento di Fisica "E.R. Caianiello", Università degli Studi di Salerno, Fisciano, SA, Italy

Chapter 1
Introduction to the Volume

Roberta Citro and Ferdinando Mancini

Abstract A brief introduction on the content of the Volume with a summary of the main topics for each Chapter is reported. The Volume covers the physics of correlated systems out of equilibrium from a theoretical, numerical and computational point of view. It also contains training sessions, which could be beneficial to researchers approaching the field of nonequilibrium many-body systems. The readers will strongly benefit from the different overviews.

1.1 Introduction

In the last decade, novel pathways to explore correlated materials have been devised, ranging from hetero- and nano-structuring to the possibility to detect measurements of the real-time evolution of samples brought out of equilibrium by an external stimulus. The latter is what happens in pump-and-probe spectroscopies [1], where a laser excites the system in a non-thermal excited state, and different spectroscopies (optical, photoemission) are performed at different delays to monitor the evolution of the system and its relaxation to equilibrium. These approaches have been used to explore a variety of properties of strongly correlated materials with a particular focus on high-temperature superconductors, addressing, the electron-boson coupling [2–4], the condensate dynamics [5], the phonon response [6, 7], the quasiparticle dynamics [8, 9], the high-energy dynamics of the pseudogap and of the Mott gap [10–12]. Last,

R. Citro (✉) · F. Mancini
IIASS, International Institute for Advanced Scientific Studies "E.R. Caianiello",
Vietri sul Mare, SA, Italy
e-mail: citro@sa.infn.it

F. Mancini
e-mail: mancini@sa.infn.it

R. Citro · F. Mancini
Dipartimento di Fisica "E.R. Caianiello",
Università degli Studi di Salerno, I-84084 Fisciano, SA, Italy

R. Citro and F. Mancini (eds.), *Out-of-Equilibrium Physics of Correlated Electron Systems*, Springer Series in Solid-State Sciences 191,
https://doi.org/10.1007/978-3-319-94956-7_1

but not least, we mention experiments where excitation of lattice degrees of freedom can induce remarkable non-equilibrium phenomena [13–15].

On the other hand, various protocols can be used to study the non-equilibrium dynamics in cold-atoms systems [16, 17]. The main difference between actual solids and cold-atom systems is that the latter can reasonably be described as closed quantum systems, while in the former the electronic fluid is coupled with the environment and it can therefore dissipate the energy accumulated during the excitation process.

The fast developments of these experimental techniques have triggered the development of theoretical methods to study correlated fermions out of equilibrium. However, since the solution of the problem of many interacting electrons is notoriously a very hard problem, we can not hope to obtain exact solutions for the time-dependent properties except for very small systems which allow for an exact diagonalization of the Hamiltonian matrix. This Volume intends to introduce the readers to some of the approaches which have been introduced to study correlated systems, extended out of equilibrium. Among them one can quote the more analytical approaches based on non-equilibrium Green's function formalism and the mean-field type of approaches. Among the most successful, one can mention the Density-Matrix Renormalization Group (DMRG) [18] and the Dynamical Mean-Field Theory (DMFT) [19]. The DMRG works particularly well in one spatial dimension, a situation which can certainly be simulated with ultracold atoms while the DMFT, which becomes exact in the limit of infinite coordination, has been shown to reasonably describe the properties of three-dimensional systems.

Chapter 1 focuses on some "mean-field" approaches who share some conceptual similarities with DMFT and they can in fact be seen as simplified versions of DMFT. Among these, are discussed methods based on slave particles (slave bosons [20–22], slave rotors [23], slave spins [24]) and the methods based on the Gutzwiller wave approximation, which is equivalent to saddle point of the slave-boson approach. In particular, the focus will be on the Gutzwiller method, which has been applied to several non-equilibrium problems in the last few years. In order to better assess the quality of the Gutzwiller solution, the Chapter also addresses the comparison with DMFT results for the same problems. Thus a brief introduction on DMFT is also present, mainly to convince the reader that the Gutzwiller approximation can be viewed as a simplified version of the more accurate DMFT.

Chapter 2 starts from the description of the real-time Greens functions, which provide a rigorous framework to interpret electronic structure out of equilibrium. Afterwards the Keldysh formalism and its relation to the description of non-equilibrium states in terms of kinetic equations is introduced. Finally the non-equilibrium DMFT theory and some of its applications are largely discussed. The focus will be on photo-induced states in Mott insulators, which provides a paradigm example for a non-equilibrium system where well-defined quasiparticles are not established.

The goal of Chap. 3 is to illustrate connections between two widely used, but often separately adopted approaches to deal with quantum systems out of equilibrium, namely quantum master equations and nonequilibrium Green's functions. In particular, the paradigmatic case of the Anderson impurity model out of equilibrium is described and its description from one approach to the other is discussed.

At the end of the chapter, is shown how the best of the master equations versus the Green's function approach can be combined to obtain a highly accurate solution of this model, which resolves the nonequilibrium Kondo physics down to temperatures well below the Kondo scale.

As a training session, this Chapter is largely devoted to an introduction to the Lindblad quantum master equation based on standard treatments, as well as methods to solve this equation.

References

1. C. Giannetti, M. Capone, D. Fausti, M. Fabrizio, F. Parmigiani, D. Mihailovic, Ultrafast optical spectroscopy of strongly correlated materials and high-temperature superconductors: a non-equilibrium approach. Adv. Phys. **65**(2), 58–238 (2016)
2. S. Dal Conte, C. Giannetti, G. Coslovich, F. Cilento, D. Bossini, T. Abebaw, F. Banfi, G. Ferrini, H. Eisaki, M. Greven, A. Damascelli, D. van der Marel, F. Parmigiani, Disentangling the electronic and phononic glue in a high-Tc superconductor. Science **335**(6076), 1600–1603 (2012)
3. M. Sentef, A.F. Kemper, B. Moritz, J.K. Freericks, Z.-X. Shen, T.P. Devereaux, Examining electron-boson coupling using time-resolved spectroscopy. Phys. Rev. X **3**, 041033 (2013)
4. S. Dal Conte, L. Vidmar, D. Golez, M. Mierzejewski, G. Soavi, S. Peli, F. Banfi, G. Ferrini, R. Comin, B.M. Ludbrook, L. Chauviere, N.D. Zhigadlo, H. Eisaki, M. Greven, S. Lupi, A. Damascelli, D. Brida, M. Capone, J. Bonca, G. Cerullo, C. Giannetti, Snapshots of the retarded interaction of charge carriers with ultrafast fluctuations in cuprates. Nat. Phys. **11**, 421–426 (2015)
5. B. Mansart, J. Lorenzana, A. Mann, A. Odeh, M. Scarongella, M. Chergui, F. Carbone, Coupling of a high-energy excitation to superconducting quasiparticles in a cuprate from coherent charge fluctuation spectroscopy. Proc. Nat. Acad. Sci. **110**(12), 4539–4544 (2013)
6. F. Novelli, G. De Filippis, V. Cataudella, M. Esposito, I. Vergara, F. Cilento, E. Sindici, A. Amaricci, C. Giannetti, D. Prabhakaran, S. Wall, A. Perucchi, S. Dal Conte, G. Cerullo, M. Capone, A. Mishchenko, M. Grüninger, N. Nagaosa, F. Parmigiani, D. Fausti, Witnessing the formation and relaxation of dressed quasi-particles in a strongly correlated electron system. Nat. Commun. **5**, 5112 (2014)
7. F. Novelli, G. Giovannetti, A. Avella, F. Cilento, L. Patthey, M. Radovic, M. Capone, F. Parmigiani, D. Fausti, Localized vibrations in superconducting $YBa_2Cu_3O_7$ revealed by ultrafast optical coherent spectroscopy. Phys. Rev. B **95**, 174524 (2017)
8. J.D. Rameau, S. Freutel, L. Rettig, I. Avigo, M. Ligges, Y. Yoshida, H. Eisaki, J. Schneeloch, R.D. Zhong, Z.J. Xu, G.D. Gu, P.D. Johnson, U. Bovensiepen, Photoinduced changes in the cuprate electronic structure revealed by femtosecond time- and angle-resolved photoemission. Phys. Rev. B **89**, 115115 (2014)
9. C.L. Smallwood, W. Zhang, T.L. Miller, G. Affeldt, K. Kurashima, C. Jozwiak, T. Noji, Y. Koike, H. Eisaki, D.-H. Lee, R.A. Kaindl, A. Lanzara, Influence of optically quenched superconductivity on quasiparticle relaxation rates in $Bi_2Sr_2CaCu_2O_{8+\delta}$. Phys. Rev. B **92**, 161102 (2015)
10. C. Giannetti, F. Cilento, S.D. Conte, G. Coslovich, G. Ferrini, H. Molegraaf, M. Raichle, R. Liang, H. Eisaki, M. Greven, A. Damascelli, D. van der Marel, F. Parmigiani, Revealing the high-energy electronic excitations underlying the onset of high-temperature superconductivity in cuprates. Nat. Commun. **2**, 353 (2011)
11. F. Cilento, S. Dal Conte, G. Coslovich, S. Peli, N. Nembrini, S. Mor, F. Banfi, G. Ferrini, H. Eisaki, M.K. Chan, C.J. Dorow, M.J. Veit, M. Greven, D. van der Marel, R. Comin, A. Damascelli, L. Rettig, U. Bovensiepen, M. Capone, C. Giannetti, F. Parmigiani, Photo-enhanced

antinodal conductivity in the pseudogap state of high-Tc cuprates. Nat. Commun. **5**, 4353 (2014)
12. S. Peli, S.D. Conte, R. Comin, N. Nembrini, A. Ronchi, P. Abrami, F. Banfi, G. Ferrini, D. Brida, S. Lupi, M. Fabrizio, A. Damascelli, M. Capone, G. Cerullo, C. Giannetti, Mottness at finite doping and charge instabilities in cuprates. Nat. Phys. **13**, 806–811 (2017)
13. D. Polli, M. Rini, S. Wall, R.W. Schoenlein, Y. Tomioka, Y. Tokura, G. Cerullo, A. Cavalleri, Coherent orbital waves in the photo-induced insulator-metal dynamics of a magnetoresistive manganite. Nat. Mater. **6**, 643 (2007)
14. D. Fausti, R.I. Tobey, N. Dean, S. Kaiser, A. Dienst, M.C. Hoffmann, S. Pyon, T. Takayama, H. Takagi, A. Cavalleri, Light-induced superconductivity in a stripe-ordered cuprate. Science **331**(6014), 189–191 (2011)
15. R. Mankowsky, A. Subedi, M. Forst, S.O. Mariager, M. Chollet, H.T. Lemke, J.S. Robinson, J.M. Glownia, M.P. Minitti, A. Frano, M. Fechner, N.A. Spaldin, T. Loew, B. Keimer, A. Georges, A. Cavalleri, Nonlinear lattice dynamics as a basis for enhanced superconductivity in $YBa_2Cu_3O_{6.5}$. Nature **516**, 71–73 (2014)
16. T. Langen, R. Geiger, J. Schmiedmayer, Ultracold atoms out of equilibrium. Ann. Rev. Condens. Matt. Phys. **6**(1), 201–217 (2015)
17. A. Polkovnikov, K. Sengupta, A. Silva, M. Vengalattore, Colloquium. Rev. Mod. Phys. **83**, 863–883 (2011)
18. U. Schollwöck, The density-matrix renormalization group. Rev. Mod. Phys. **77**, 259–315 (2005)
19. A. Georges, G. Kotliar, W. Krauth, M.J. Rozenberg, Dynamical mean-field theory of strongly correlated fermion systems and the limit of infinite dimensions. Rev. Mod. Phys. **68**, 13–125 (1996)
20. G. Kotliar, A.E. Ruckenstein, New functional integral approach to strongly correlated fermi systems: the gutzwiller approximation as a saddle point. Phys. Rev. Lett. **57**, 1362–1365 (1986)
21. F. Lechermann, A. Georges, G. Kotliar, O. Parcollet, Rotationally invariant slave-boson formalism and momentum dependence of the quasiparticle weight. Phys. Rev. B **76**, 155102 (2007)
22. A. Isidori, M. Capone, Rotationally invariant slave bosons for strongly correlated superconductors. Phys. Rev. B **80**, 115120 (2009)
23. S. Florens, A. Georges, Slave-rotor mean-field theories of strongly correlated systems and the mott transition in finite dimensions. Phys. Rev. B **70**, 035114 (2004)
24. L. de' Medici, M. Capone, *Modeling Many-Body Physics with Slave-Spin Mean-Field: Mott and Hund's Physics in Fe-Superconductors* (Springer International Publishing, Cham, 2017), pp. 115–185

Chapter 2
Towards the Understanding of Superconductors and Correlated Materials out of Equilibrium: Mean Field Approaches

Massimo Capone and Carla Lupo

Abstract Lectures prepared for the XX Training Course in the Physics of Strongly Correlated Systems held in Vietri sul Mare (Sa), October 3–7, 2016.

2.1 Motivation and Introduction

In the last thirty years the field of strongly correlated electron systems has established as one of the most active and lively, but sometimes frustrating fields of condensed-matter physics. Needless to say, the surge of the field has been motivated by the discovery of high-temperature superconductivity in doped copper oxides, but the following decades have seen the rise of a number of materials and artificial solids demonstrating that the interaction between particles can lead to a virtually infinite landscape of novel quantum phases and of spectacular realizations of more traditional phenomena.

In the last few years, novel pathways to explore these materials have been devised, ranging from hetero- and nano-structuring to the possibility to detect measure the real-time evolution of a correlated material brought out of equilibrium by an external stimulus. The latter is, in a nutshell, what happens in pump-and-probe spectroscopies [1], where a laser excites the system in a non-thermal excited state, and different spectroscopies (optical, photoemission) are performed at different delays to monitor the evolution of the system and its relaxation to equilibrium. These approaches have been used to explore a variety of properties of strongly correlated materials with a particular focus on high-temperature superconductors, addressing, the electron-boson coupling [2–4], the condensate dynamics [5], the phonon response [6, 7], the

M. Capone (✉)
Scuola Internazionale Superiore di Studi Avanzati (SISSA)
and CNR-IOM Democritos National Simulation Center, Via Bonomea 265,
34136 Trieste, Italy
e-mail: capone@sissa.it

C. Lupo
Physics King's College, London, UK

R. Citro and F. Mancini (eds.), *Out-of-Equilibrium Physics of Correlated Electron Systems*, Springer Series in Solid-State Sciences 191,
https://doi.org/10.1007/978-3-319-94956-7_2

quasiparticle dynamics [8, 9], the high-energy dynamics of the pseudogap and of the Mott gap [10–12]. Last, but not least, we mention experiments where excitation of lattice degrees of freedom can induce remarkable non-equilibrium phenomena [13–15].

On the other hand, various protocols can be used to study the non-equilibrium dynamics in cold-atoms systems [16, 17]. The main difference between actual solids and cold-atom systems is that the latter can reasonably described as open quantum systems, while in the former the electronic fluid is coupled with the environment and it can therefore dissipate the energy accumulated during the excitation process.

This short manuscript is not meant to review the variety of experiments which have been carried out in the last few years using different pump-probe set-ups to explore the non-equilibrium properties of materials. For a recent review we suggest [1]. Our point of view is that the non-equilibrium properties of correlated materials and high-temperature superconductors should not be viewed as a novel, extremely hard, challenge to be added to those raised by the equilibrium properties, but rather as a novel knob that we can use to understand the elusive properties of these materials.

The fast developments of these experimental techniques have triggered the development of theoretical methods to study correlated fermions out of equilibrium. However, since the solution of the problem of many interacting electrons is notoriously a very hard problem, we can not hope to obtain exact solutions for the time-dependent properties except for very small systems which allow for an exact diagonalization of the Hamiltonian matrix. Among the many approaches that have been introduced to study correlated systems, some have already been extended out of equiibrium. Among the most successful, we mention the Density-Matrix Renormalization Group (DMRG) [18] and the Dynamical Mean-Field Theory (DMFT) [19]. We will not discuss here the DMRG, which is constructed to work remarkably well in one spatial dimension, a situation which can certainly be simulated with ultracold atoms, but it is approximately relevant only to a small set of materials. On the other hand, the DMFT, which becomes exact in the limit of infinite coordination, has been shown to reasonably describe the properties of three-dimensional systems.

The formalism and applications of non-equilibrium DMFT approach have been reviewed in [20] and are discussed in another contribution in the present book. In this chapter we focus on some "mean-field" approaches who share some conceptual similarities with DMFT and they can in fact be seen as simplified versions of DMFT. Among these, we refer to the methods based on slave particles (slave bosons [21–23], slave rotors [24], slave spins [25]) and the method based on the Gutzwiller wave approximation, which is equivalent to the saddle point of the slave-boson approach. For the sake of definiteness, we will in fact focus on the Gutzwiller method, which has been applied to several non-equilibrium problems in the last few years. In order to better assess the quality of the Gutzwiller solution, we will compare with DMFT results for the same problems. For this reason we will also briefly introduce DMFT, mainly to convince the reader that the Gutzwiller approximation can be viewed as a simplified version of the more accurate DMFT.

Strongly Correlated Electrons: A Very Short Introduction

The quantum theory of the electronic properties of solids is based on the so-called band theory of solids, in which the electronic structure determined by the crystal structure is constituted by a series of energy bands possibly separated by forbidden energy regions (gaps). The levels are populated by the electrons fulfilling the Pauli principle. A system is a metal if some band is partially filled, while it is an insulator only if the number of electrons is such to completely fill a band, leaving the next band completely empty.

Within this theory, the state of any given electron is completely unaffected by the presence of the others. If one more electron is added in a system, the other electrons persist in their state ignoring the new one. In other words, the band theory of solids describes a set of non-interacting electrons and it is what we define a single-particle approach. This can be quite surprising, because the electrons are strongly interacting through the Coulomb repulsion (and also via the interaction with the lattice). The reason why the band theory works for a large number of compounds is mainly the screening of the Coulomb interaction.

The Landau theory of Fermi Liquids is a phenomenological description of an interacting systems in terms of weakly interacting "quasiparticles". More precisely, at low energy and low temperatures, the spectrum of excitation of a "liquid" of interacting fermions is in one-to-one correspondence with that of a non-interacting system. These excitations are called quasiparticles and they have fermionic character. If compared with the original fermions, the quasiparticles have an effective mass larger than the bare band mass due to the interactions and they are weakly interacting. Therefore, most observables have a behavior similar to a gas of non-interacting electrons, but with renormalized parameters.

The Landau approach breaks down when the interactions are sufficient to drive a phase transition to an insulator or possibly to novel metallic states with anomalous properties (non-Fermi liquid metals). This happens in materials with open d-or f-electron shells, whose orbitals are localized and the bands are narrow, i.e., they span a small region of energy: in these conditions the effect of the screened Coulomb interaction can not be neglected and the mutual influence between electrons is too pronounced that mean field theories can not be applied. The most spectacular breakdown of the Landau theory has been first discovered already in 1937 when de Boer Verwey observed that V_2O_3 and other transition-metal oxides are insulators despite the count of the electrons would imply partially-filled bands and a metallic beavior. This kind of insulating state is usually called a Mott insulator. Even before the complete breakdown, narrow-band electrons can show anomalies in their responses and phenomena which are incompatible with a picture of independent electrons. It is widely believed that such anomalous phases are strongly intertwined with the most fascinating phenomena displayed by correlated-electron materials, including high-temperature superconductivity.

The correlation-driven insulators are called Mott insulators because Sir Nevill Mott [26] has been indeed the first to realize that the insulating behavior of transition metal oxides was the result of the interaction between the electrons and to propose that a metal-insulator transition could be the result of the competition between the kinetic energy and the correlation which constraints the motion of the electrons. In such insulators, the lack of electronic conductions is not due to the lack of low-energy states, but it is associated to a localization of the carriers bound to atoms with open shells. As opposed to metals and band insulators, this electronic state is well described by a real-space picture of almost localized electrons which move only incoherently.

In the following we will discuss several phenomena connected with Mott insulators and with the Mott transition, namely the transition between a metal and an insulator driven by some control parameter like the change of the electronic bandwidth (for example by applying pressure or chemical substitution) or the doping. The reason for the focus on Mott physics is that the Mott localization is the clearest and most direct signature of strong correlation physics, and its understanding is instrumental to characterize the other consequences of this physics.

As we will mention in the following, the proximity to a Mott transition is a very fertile ground for the appearance of novel physics and/or of spectacular and surprising realization of known phenomena. Among these, high-temperature superconductivity is undoubtedly the most significant and popular, but many other examples can be given, ranging from colossal magnetoresistance to non-Fermi liquid metals, to anomalous phases with spin and orbital ordering. All these phases can be realized by applying external handles like pressure, doping or magnetic field to Mott insulators. The main reason for this richness is that highly correlated metallic phases are characterized by a small kinetic energy due to the presence of correlations. As a result, the metallic state is fragile with respect to almost every external field, and it becomes unstable towards the different phases we mentioned. Therefore the understanding of how a metal becomes unstable towards a Mott insulator is a necessary step for the understanding of all the exciting phases we can encounter close to Mott localization.

2.2 The Gutzwiller Variational Method

In this chapter we introduce the variational method based on the pioneering work by Martin Gutzwiller in 1964 [34, 35]. The Gutzwiller approach is an application of the variational principle in which a simple variational wavefunction is introduced to describe the effects of the interactions in a Hubbard model. As we shall discuss in the following, what goes under the name of Gutzwiller approach requires a further approximation on top of the variational ansatz. The original approximation introduced ad hoc by Gutzwiller in order to obtain an analytical result has been later shown to provide the exact expectation value of the Hamiltonian on the Gutzwiller wave function in the limit of infinite coordination (or dimensionality). This also establishes a link with Dynamical Mean-Field Theory, which is indeed exact in the same limit of infinite coordination.

As we shall se, the Gutzwiller approximation is in a sense a non-perturbative method to study strongly correlated electrons, which makes it particularly useful to study the Mott-Hubbard transition that we briefly introduced in the previous chapter. For historical reasons the Gutzwiller picture of the Mott transition is usually called

Brinkman and Rice transition [36]. In this section we start by recalling the basic aspects of the equilibrium Gutzwiller approximation which are crucial to understand the non-equilibrium scenario.

The Gutzwiller approximation is also completely equivalent to the popular slave-boson mean-field approach, in which auxiliary bosonic particles are introduced to describe the high-energy physics. In particular, the version of the Gutzwiller approximation that we describe in these notes is equivalent to the so-called rotationally invariant slave bosons, which have been introduced in [22] and later generalized to the superconducting state in [23]. The historical Gutziller approach is instead completely equivalent to the Kotliar-Ruckenstein formulation of the slave boson mean-field.

2.2.1 *The Time-Independent Gutzwiller Approximation*

In order to clarify the physical idea behind the Gutzwiller approximation we start introducing the original formulation proposed by M. C. Gutzwiller for the single-band Hubbard model, even if the remainder of these notes will be based on a more general formulation introduced recently, which can be used for more complicated multiband models and it is readily generalized out of equilibrium.

The starting point of the method is the introduction of a variational ansatz for a wave function which introduces the effect of a Hubbard-like interaction term starting from a non-interacting wavefunction described by a Slater determinant. In the case of the single-band Hubbard model we can start from

$$|\Psi_G\rangle = g^{\hat{D}}|FS\rangle \tag{2.1}$$

where $|FS\rangle = \prod_{k<k_F,\sigma} c^\dagger_{k,\sigma}|0\rangle$ is a Slater determinant that we take as the non interacting Fermi sea, which is the exact solution of the Hubbard model in the absence of interactions. $\hat{D} = \sum_i n_{i\uparrow}n_{i\downarrow}$ is an operator whose expectation value D is the number of the double occupied sites in a given configuration, and g is a variational parameter. Our task is to minimize the expectation value of the Hamiltonian on the trial wavefunction as a function of the variational parameter g.

The idea behind this wavefunction is to introduce in a variational way a "real-space perspective" starting from a wavefunction which is diagonal in momentum space, trying to overcome the duality between interaction and kinetic energy that makes the model hard to solve. Obviously $g = 1$ for $U = 0$ where we recover the Fermi sea, while the limit of infinite U corresponds to $g \to 0$. Therefore we expect that increasing U, g will evolve from 1 to 0. As a matter of fact, the rationale of the Gutzwiller wavefunction is to give a reduced weight to configurations which have a larger value of double occupancy D. The larger the value of the interaction, the smaller the weight. Even if the Fermi sea is easily written in momentum space, we may imagine to expand it onto a real-space basis and we can formally write

$$|\Psi_G\rangle = \sum_D \sum_{\{i_D\}} A_{i_D} |\Psi_{i_D}\rangle, \tag{2.2}$$

where D are the different eigenvalues of the double-occupancy operator $\hat{D}$, $|\Psi_{i_D}\rangle$ are all the basis states in the Fock space with D double occupancies, spanned by the index i_D. A_{i_D} is the coefficient of the expansion of the Fermi sea in the basis of local configurations. In principle A_{i_D} is known, but it is in practice impossible to use this information to compute the expectation values of the Hamiltonian. For this reason Gutzwiller assumed that all the matrix elements were independent on the actual form of the state $|\Psi_{i_D}\rangle$, but they only depended on the number of double occupancies D. This is, in a nutshell, the original formulation of the Gutzwiller approximation. In this approximation, the only effect of U is to reduce the number of doubly occupied sites, thereby leading to a renormalization of the kinetic energy.

It is useful to stress what is the physics that the Gutwiller approximation overlooks. Let us consider for example the two configurations A and B for a four-site model.

$$\mathrm{A}: \ |\uparrow\downarrow, \uparrow\downarrow, \uparrow, 0\rangle \qquad \mathrm{B}: \ |\uparrow\downarrow, 0, \uparrow\downarrow, \uparrow\rangle \tag{2.3}$$

A and B have the same number of doubly occupied sites, therefore they will have the same weight in the Gutzwiller approximation. On the other hand, if we apply the hopping Hamiltonian to B, we can perform more hopping processes than in the case of A (the key are the two neighboring double occupancies in A). In other words, the Gutzwiller approximation completely neglects the spatial arrangement of the local configurations, focusing only on the frequency of the different local configurations. In this regard, one can immediately see the link with the limit of infinite dimensionality. In this limit, the details of all the sites connected with an arbitrary one are lost in an average mean-field.

After we have introduced the main concepts behind the Gutzwiller approximation, we can move to a more recent and general formalism, which includes the possibility to treat multiple orbitals with arbitrary on-site interactions and hybridizations. The general form of the Gutzwiller ansatz can be written simply as

$$|\Psi_{\mathrm{G}}\rangle = \prod_{\mathbf{R}} \mathscr{P}_{\mathbf{R}} |\Psi_0\rangle \tag{2.4}$$

with $|\Psi_0\rangle$ being the one-body wave function and $P_{\mathbf{R}}$ are projectors operators acting on the local Hilbert space, which generalize the single partial projection operator g^D. The aim of the method is to find, within the class of the wave-functions defined above, the best approximation to the true ground state, minimizing the variational energy

$$E_{\mathrm{G}} = \frac{\langle \Psi_{\mathrm{G}} | \mathscr{H} | \Psi_{\mathrm{G}} \rangle}{\langle \Psi_{\mathrm{G}} | \Psi_{\mathrm{G}} \rangle}. \tag{2.5}$$

The exact calculation of this expectation value can only be carried out numerically for finite-size systems, while a semi-analytical treatment is possible in the limit of infinite coordination number $z \to \infty$ which, as we mentioned, amounts to a kind of spatial mean-field, as long as the two following constraints are satisfied:

$$\langle \Psi_0 | \mathscr{P}_{\mathbf{R}}^{\dagger} \mathscr{P}_{\mathbf{R}} | \Psi_0 \rangle = 1 \tag{2.6}$$

$$\langle \Psi_0 | \mathscr{P}_{\mathbf{R}}^{\dagger} \mathscr{P}_{\mathbf{R}}\, c_{\mathbf{R}a}^{\dagger}\, c_{\mathbf{R}b} | \Psi_0 \rangle = \langle \Psi_0 | c_{\mathbf{R}a}^{\dagger}\, c_{\mathbf{R}b} | \Psi_0 \rangle \tag{2.7}$$

Furthermore, since the wave-function is normalized, given any local operator $\mathscr{O}_{\mathbf{R}}$, we can obtain an explicit expression of its expectation value

$$\langle \Psi_0 | \mathscr{P}^{\dagger} \mathscr{O}_{\mathbf{R}} P | \Psi_0 \rangle = \langle \Psi_0 | P_{\mathbf{R}}^{\dagger} \mathscr{O}_{\mathbf{R}} P_{\mathbf{R}} | \Psi_0 \rangle \tag{2.8}$$

which means that in computing averages of local operators, only the local projectors play a role.
Similarly, the expectation value of the hopping operator reads

$$\langle \Psi_0 | \mathscr{P}^{\dagger}\, c_{ia}^{\dagger}\, c_{jb} \mathscr{P}\, | \Psi_0 \rangle = \langle \Psi_0 | \mathscr{P}_i^{\dagger}\, c_{ia}^{\dagger}\, \mathscr{P}_i\, \mathscr{P}_j^{\dagger}\, c_{jb}\, P_j\, | \Psi_0 \rangle \tag{2.9}$$

$$= \textstyle\sum_{cd} \mathscr{R}_{iac}^{*}\, \mathscr{R}_{jbd} \langle \Psi_0 | c_{ic}^{\dagger} c_{jd} | \Psi_0 \rangle \tag{2.10}$$

where $\mathscr{R}_{jbd}$ is defined from the equation

$$\langle \Psi_0 | \mathscr{P}^{\dagger}\, c_{ia}^{\dagger}\, c_{ib} \mathscr{P}\, | \Psi_0 \rangle = \sum_c \mathscr{R}_{iac}^{*} \langle \Psi_0 | c_{ic}^{\dagger}\, c_{id} | \Psi_0 \rangle \tag{2.11}$$

The main consequence of the introduction of $\hat{\mathscr{R}}_i$ is that it defines a new one-body effective Hamiltonian $\mathscr{H}^*$ from the non-interacting part of the original Hamiltonian with renormalized hopping amplitudes $\hat{t}_{ij}^{*ab}$

$$\mathscr{H}^* = \sum_{ij} t_{ij}^{*ab}\, c_{ia}^{\dagger}\, c_{jb} \qquad t_{ij}^{*ab} = \sum_{cd} R_{ica}^{*} R_{jdb} t_{ij}^{cd} \tag{2.12}$$

$\mathscr{H}^*$ can obviously be diagonalized in momentum space, leading to itinerant eigenfunctions. Therefore it describes metallic states for any non-trivial density. These renormalized single-particle states are a realization of the concept of Landau quasi-particles, and they are characterized by an effective mass enhancement given by the electron-electron correlations. We can expect that the renormalization factors become smaller and smaller as the interaction grows, leading to a reduced mobility of the carriers. As we shall see, this is the way in which the destruction of the metallic state is described in the Gutzwiller approximation.

Starting from (2.5), we can now procede in minimizing with respect to the projectors as follows:

$$E = \min_{|\Psi_0\rangle, \mathcal{P}} \left[\langle \Psi_0 | \mathcal{H}^* | \Psi_0 \rangle + \langle \Psi_0 | P_i^\dagger \mathcal{H}_i P_i | \Psi_0 \rangle \right] \tag{2.13}$$

In order to simplify the expression of the expected values introduced before, it is convenient to introduce a specific representation. We will now define a natural basis which diagonalizes the density matrix operator in the slater determinant.

$$\langle \Psi_0 | d_{ia}^\dagger d_{ib} | \Psi_0 \rangle = \delta_{ab} \, n^{(0)_{ia}} \tag{2.14}$$

In the case of a single-orbital Hubbard model (and of any multi-orbital model with purely density-density interactions) the natural basis coincides with the original orbital basis, since the density matrix is already diagonal and the following construction is obviously redundant.

In the most general case the basis of operators for the natural basis is related to the original fermionic operators via a unitary transformation. We can introduce Fock states in both bases

$$|i, n\rangle = \prod_a \left(d_{ia}^\dagger \right)^{n_a} |0\rangle \tag{2.15}$$

$$|i, \Gamma\rangle = \prod_a \left(c_{ia}^\dagger \right)^{\Gamma_a} |0\rangle \tag{2.16}$$

In the new basis the matrix of the local occupation probability is diagonal:

$$P_{i,n}^{(0)} = \langle \Psi_0 | \, i, \, n \rangle \langle i, n | \Psi_0 \rangle = \prod_a \left(n_a^{(0)} \right)^{n_a} \left(1 - n_a^{(0)} \right)^{1-n_a} \tag{2.17}$$

We can now parametrize the Gutzwiller projectors in a mixed original/natural basis representation:

$$P_i = \sum_{\Gamma, n} \frac{\Phi_{i,\Gamma,n}}{\sqrt{P_{i,n}^{(0)}}} |i, \Gamma\rangle \langle i, n| \tag{2.18}$$

where the variational parameters $\Phi_{i;\Gamma,n}$ define a local variational matrix $\hat{\Phi}_i$ and $|i; \Gamma\rangle$ are basis set in the terms of the original operators $c_{ia}^\dagger$. The main results of this manipulation is that at this point instead of minimizing with respect to the projectors $\mathcal{P}$, we perform the minimization w.r.t. $\Phi_{i,n,\Gamma}$. Furthermore under this basis transformation the matrix representation of the creation/annihilation operators becomes

$$\left(\hat{d}_{ia}^\dagger \right)_{n_1, n_2} = \langle i, n_1 | d_{ia}^\dagger | i, n_2 \rangle \tag{2.19}$$

$$\left(\hat{c}_{ia}^\dagger \right)_{\Gamma_1, \Gamma_2} = \langle i, \Gamma_1 | c_{ia}^\dagger | i, \Gamma_2 \rangle \tag{2.20}$$

and the constraints previously introduced in (2.6)

$$\mathrm{Tr}\left[\Phi_i^\dagger \Phi_i\right] = 1 \tag{2.21}$$

$$\mathrm{Tr}\left[\Phi_i^\dagger \Phi_i \, d_{ia}^\dagger \, d_{ib}\right] = \delta_{ab}\langle\Psi_0|d_{ia}^\dagger \, d_{ib}|\Psi_0\rangle = \delta_{a,b} n_{ia}^{(0)} \tag{2.22}$$

The variational energy to minimize becomes:

$$E = \min_{|\Psi_0\rangle,\Phi}\left[\sum_{ij}\left(\sum_{ab} t_{ij}^{*ab}\langle\Psi_0|d_{ia}^\dagger d_{jb}|\Psi_0\rangle\right) + \mathrm{Tr}\left[\Phi_i^\dagger \mathscr{H}_i \Phi_i\right]\right] \tag{2.23}$$

where the effective hopping matrix elements are given by

$$t_{ij}^{*ab} = \sum_{cd} R_{ica}^* R_{jdb}\, t_{ij}^{cd} \tag{2.24}$$

and

$$R_{\mathbf{R}ab}^* = \frac{1}{\sqrt{n_{\mathbf{R}b}^{(0)}(1-n_{\mathbf{R}b})^{(0)}}}\mathrm{Tr}\left[\hat{\Phi}_{\mathbf{R}}^\dagger \hat{c}_{\mathbf{R}a}^\dagger \hat{\Phi}_{\mathbf{R}} \hat{d}_{\mathbf{R}b}\right] \tag{2.25}$$

where we remind that $\hat{c}^\dagger$, $\hat{d}^\dagger$ ae the creation operators on different basis and $\hat{\Phi}$ is an unknown matrix.

2.2.2 An Explicit Example: The Single Band Hubbard Model

The derivation we have just given shows that, within the Gutzwiller approximation, the groundstate energy of an interacting model can be mapped onto an effective single particle Hamiltonian with renormalized hoppings which include the effect of the interactions.

In order to understand how the Gutzwiller approximation works, its successes and limitations, we now briefly discuss the application to the half-filled single-band hubbard model in equilibrium and to the Mott-Hubbard transition. This discussion will also be the basis of the non-equilibrium implementation that we will discuss at length in the following.

We consider the single-band (or equivalently single-orbital) Hubbard Hamiltonian. The local Fock states are obviously

$$|0\rangle,\ |\uparrow\rangle,\ |\downarrow\rangle,\ |\uparrow\downarrow\rangle \tag{2.26}$$

and the local density matrix is already diagonal in this basis. Therefore we do not need to distinguish between the original basis and the natural orbital basis, which means that the d fermions coincide with the c fermions. In this basis every local operator can be written as a 4 × 4 matrix. For example the creation and annihilation operators read

$$\hat{c}^{\dagger}_{\uparrow} = \begin{pmatrix} 0\,1\,0\,0 \\ 0\,0\,0\,0 \\ 0\,0\,0\,1 \\ 0\,0\,0\,0 \end{pmatrix} \qquad \hat{c}_{\uparrow} = \begin{pmatrix} 0\,0\,0\,0 \\ 1\,0\,0\,0 \\ 0\,0\,0\,0 \\ 0\,0\,1\,0 \end{pmatrix} \tag{2.27}$$

Because the density matrix is diagonal, the most general variational matrix $\hat{\Phi}_i$ which satisfies the Gutzwiller constraints is:

$$\hat{\Phi}_i = \begin{pmatrix} \Phi_{i0} & 0 & 0 & 0 \\ 0 & \Phi_{i\uparrow} & 0 & 0 \\ 0 & 0 & \Phi_{i\downarrow} & 0 \\ 0 & 0 & 0 & \Phi_{i\downarrow\uparrow} \end{pmatrix} \tag{2.28}$$

If we consider the half-filled Hubbard model with nearest-neighbor hopping only, the model has particle-hole symmetry, which implies $\Phi_{i0} = \Phi_{i\uparrow\downarrow} = \Phi_0$. If we limit ourselves to nonmagnetic or paramagnetic solutions $\Phi_{i\uparrow} = \Phi_{i\downarrow} = \Phi_1$. If we further restrict to homogeneous solutions where the variational parameters to not depend on the position on the lattice, also the hopping renormalization factor is site- and spin-independent and reads

$$R = 2\left[\Phi_0^* \Phi_1^* + \Phi_1^* \Phi_0\right] \tag{2.29}$$

Using the normalization condition $2\left(|\Phi_0|^2 + |\Phi_1|^2\right) = 1$ we can get rid of Φ_1 and we write the variational energy as a function of the single variational parameter Φ_0.

$$E = \min_{|\psi\rangle, \Phi_0^2} \left[-R^2(\Phi_0) t \sum_{\langle i,j\rangle,\sigma} \langle \psi_0 | c^{\dagger}_{i\sigma} c_{j\sigma} | \psi_0 \rangle + U \sum_i \Phi_0^2 \right] \tag{2.30}$$

where the interaction term comes from

$$\mathrm{Tr}\left[\Phi_i^{\dagger} \mathscr{H}_i \Phi_i\right] \quad \text{and} \quad \mathscr{H}_i = U \begin{pmatrix} 0\,0\,0\,0 \\ 0\,0\,0\,0 \\ 0\,0\,0\,0 \\ 0\,0\,0\,1 \end{pmatrix} \tag{2.31}$$

Here the $|\psi_0\rangle$ is the non interacting Fermi sea. We notice that the energy only depends on the single variational parameter Φ_0, which we can rename $\Phi_0^2 = D$ because it coincides with the expectation value of the double occupancy operator. This shows also that the approach we described coincides, for a single-band model, with the original formulation by Gutzwiller.

We are therefore left with a simple and transparent formula for the energy and its dependence on the variational parameter D

$$E(D) = -N_s 8D(1-2D)\bar{\varepsilon} + UDN_s \tag{2.32}$$

being N_s the number of sites and $\bar{\varepsilon}$ is the average hopping energy defined as

$$\bar{\varepsilon} = \frac{t}{N_s} \sum_{\langle i,j\rangle,\sigma} \langle \psi_0 | c^\dagger_{i\sigma} c_{j\sigma} | \psi_0 \rangle + H.c. \tag{2.33}$$

Minimizing with respect to D, we obtain

$$\begin{cases} D = \frac{1}{4}\left(1 - \frac{U}{U_c}\right) & U < U_c, \\ D = 0 & U \geq 8\bar{\varepsilon} \end{cases} \tag{2.34}$$

where $U_c \equiv 8\bar{\varepsilon}$ is defined as critical value of the interaction above which the number of doubly occupied sites vanishes, as expected in the extreme limit of a Mott insulator. Plugging (2.34) into (2.36) we obtain

$$E = -N_s\bar{\varepsilon}\left(1 - \frac{U}{U_c}\right)^2, \tag{2.35}$$

while the kinetic energy renormalization reads

$$Z = 1 - \frac{U^2}{U_c^2} \tag{2.36}$$

These results show that the Gutzwiller approximation describes a metal-insulator transition. For $U < U_c$ the result is a metallic solution with a renormalized kinetic energy, a reduced number of double occupancies and an increasing energy as a function of U. All the quantities are reduced monotonically in modulus as U is increased and vanish at U_c. In particular the energy vanishes quadratically so that the transition is of second order. On the other hand, the picture of the insulator is trivial: For $U > U_c$ the double occupancy and the kinetic energy are always zero. Interestingly, also the response functions in the metallic solution meet the expectations for a correlated metal. In particular the charge compressibility vanishes for $U \to U_c$ as expected if a Mott gap opens, and the spin susceptibility diverges, signaling the formation of local magnetic moments.

Therefore, while the Gutzwiller approximation makes a reasonable job in describing the metallic state and its weakening and dispappearance as a function of the interaction, it does not provide a physically interesting picture of the Mott insulator. In equilibrium it would be then necessary to go beyond this approximation introducing some kind of fluctuations around this solution, but we will see that out of equilibrium the picture is richer already at the mean-field level, where the insulator acquires a non-trivial dynamics.

2.2.3 Time-Dependent Gutzwiller Approximation

In this section we move to the time-dependent Gutzwiller approximation, which allows to study non-equilibrium problems where the system can be driven out of its equilibrium state using different protocols. In this work we will mainly focus on the quantum quench protocol, and we will compare the results with time-dependent Dynamical Mean-Field Theory considering both the paramagnetic and the antiferromagnetic phases of the Hubbard model. The approach has been introduced in [37] and more details can be found in [38, 39]. We also suggest [40, 41] for detailed discussions of the method and of some of the aspects that we will cover in this manuscript. A time-dependent Gutzwiller approach was introduced in [42] but it was limited to linear response [43].

As we shall see explicitly in the following, the derivation of the time-dependent Gutzwiller approximation follows closely that of its equilibrium version. In contrast with the equilibrium case, here the problem we would like to solve is the time-dependent Schrödinger equation to find the wave function which describes the state of the system under the effect of a time-dependent Hamiltonian

$$i\partial_t|\Psi(t)\rangle = \mathcal{H}(t)|\Psi(t)\rangle. \tag{2.37}$$

Also in this case we resort to a variational principle selecting a class of wavefunctions completely equivalent to the equilibrium Gutzwiller ansatz (2.4)

$$|\Psi_G(t)\rangle = \prod_{\mathbf{R}} \mathcal{P}_{\mathbf{R}}(t)|\Psi_0(t)\rangle \tag{2.38}$$

In this case both the projectors and the Slater determinant explicitly depend on time. We introduce the action functional

$$\mathcal{S}[|\Psi\rangle] = \int_0^t d\tau\, \langle\Psi(\tau)|i\partial_\tau - \mathcal{H}(\tau)|\Psi(\tau)\rangle \tag{2.39}$$

for which a variational principle holds. Therefore, we can look for the time-dependent wavefunction $|\Psi(t)_G\rangle$ described by the Gutzwiller ansatz which makes the action $\mathcal{S}[|\Psi_G S\rangle]$ stationary.

$$\frac{\delta\mathcal{S}[|\Psi_G\rangle]}{\delta\langle\Psi_G|} = 0 \tag{2.40}$$

Also in this case we can introduce a Gutzwiller approximation which becomes exact in the limit of infinite coordination as long as the same two constraints hold. Obviously in this case the constraint must be satisfied for all the times in the time evolution

$$\langle\Psi_0(t)|\mathcal{P}^\dagger_{\mathbf{R}}(t)\mathcal{P}_{\mathbf{R}}(t)|\Psi_0(t)\rangle = 1 \tag{2.41}$$

$$\langle\Psi_0(t)|\mathcal{P}^\dagger_{\mathbf{R}}(t)\mathcal{P}_{\mathbf{R}}(t)\, c^\dagger_{\mathbf{R}a}(t)\, c_{\mathbf{R}b}(t)|\Psi_0(t)\rangle = \langle\Psi_0(t)|c^\dagger_{\mathbf{R}a}(t)\, c_{\mathbf{R}b}(t)|\Psi_0(t)\rangle \tag{2.42}$$

Given the form of the Gutzwiller wave function, the action becomes

$$\mathscr{S}_G[|\Psi\rangle] = \int_0^{t_f} d\tau \left[i\langle\psi(\tau)|\partial_\tau\psi(\tau)\rangle + i\sum_{\mathbf{R}}\langle\psi(\tau)|\mathscr{P}^\dagger_{\mathbf{R}}(\tau)\partial_\tau\left(\mathscr{P}_{\mathbf{R}}(\tau)\right)|\psi(\tau)\rangle - E(\tau)\right]$$
$$= \int_0^t d\tau\, \mathscr{L}(\tau) \tag{2.43}$$

and the Lagrangian

$$\mathscr{L}(t) = i\sum_{\mathbf{R}}\langle\Psi(t)|\mathscr{P}^\dagger_{\mathbf{R}}(t)\dot{\mathscr{P}}_{\mathbf{R}}(t)|\Psi(t)\rangle + i\langle\Psi(t)|\dot{\Psi}(t)\rangle - E(t) \tag{2.44}$$

with $E(\tau) = \langle\psi(\tau)|\mathscr{P}^\dagger(t)\mathscr{H}(t)\mathscr{P}(t)|\psi(t)\rangle$ being the total energy.

In particular, following the same scheme of the equilibrium case, the projectors can be expressed by the following equation

$$\mathscr{P}_{\mathbf{R}}(t) = \sum_{\Gamma,n}\frac{\Phi_{\mathbf{R};\Gamma,n}(t)}{\sqrt{P^{(0)}_{\mathbf{R};n}(t)}}|\mathbf{R};\Gamma\rangle\langle\mathbf{R};n| \tag{2.45}$$

where both the variational matrix $\hat{\Psi}_{\mathbf{R}}(t)$ and the local uncorrelated probability $P^{(0)}_{\mathbf{R};n}(t)$ are now time dependent:

$$P^{(0)}_{\mathbf{R};n}(t) = \langle\Psi(t)|\mathbf{R};n\rangle\langle\mathbf{R};n|\Psi(t)\rangle \tag{2.46}$$

$$\langle\Psi(t)|d^\dagger_{\mathbf{R}a}d_{\mathbf{R}b}|\Psi(t)\rangle = \delta_{ab}n^{(0)}_{\mathbf{R}a}(t) \tag{2.47}$$

Thus the explicit form for the energy becomes

$$E(t) = \langle\Psi(\tau)|\mathscr{H}_*(\tau)|\Psi(t)\rangle + \sum_{\mathbf{R}}\langle\Psi(t)|\mathscr{P}^\dagger_{\mathbf{R}}(t)H_{\mathbf{R}}\mathscr{P}_{\mathbf{R}}(t)|\Psi(t)\rangle \tag{2.48}$$

$$= \langle\Psi(t)|\mathscr{H}_*(t)|\Psi(t)\rangle + \sum_{\mathbf{R}}\mathrm{Tr}\left(\hat{\Phi}^\dagger_{\mathbf{R}}(t)\hat{\Phi}_{\mathbf{R}}(t)\hat{\mathscr{H}}_{\mathbf{R}}\right) \tag{2.49}$$

Under these assumptions the action acquires the following matrix form

$$\mathscr{S}[|\Psi\rangle] = \int_0^{t_f} d\tau \Big[i\langle\Psi(\tau)|\partial_\tau\Psi(\tau)\rangle - \langle\Psi(\tau)|\mathscr{H}_*(\tau)|\Psi(\tau)\rangle +$$
$$+ i\sum_{\mathbf{R}}\mathrm{Tr}\left(\hat{\Phi}^\dagger_{\mathbf{R}}(\tau)\partial_\tau\hat{\Phi}_{\mathbf{R}}(\tau)\right) - \sum_{\mathbf{R}}\mathrm{Tr}\left(\hat{\Phi}^\dagger_{\mathbf{R}}(\tau)\hat{\mathscr{H}}_{\mathbf{R}}\hat{\Phi}_{\mathbf{R}}(\tau)\right)\Big] \tag{2.50}$$

where $\mathscr{H}_*$ is a renormalized single-particle effective Hamiltonian whose dynamics is given by

$$i\partial_t|\Psi(t)\rangle = \mathscr{H}_*[\hat{\Phi}(t)]|\Psi(t)\rangle \tag{2.51}$$

which means that its dependence on time comes from $\hat{R}_i(t)$ according to the definition:

$$\mathcal{H}_*[\hat{\Phi}(t)] = \sum_{ij}\sum_{ab} t_{*i,j}^{a,b} d_{ia}^\dagger d_{jb} \quad \text{with} \quad t_{*i,j}^{a,b} = \sum_{\alpha\beta} R_{ia\alpha}^\dagger[\hat{\Phi}(t)] t_{i,j}^{\alpha\beta} R_{j\beta b}[\hat{\Phi}(t)] \tag{2.52}$$

and the renormalization factor is defined by the following equation

$$R_{\mathbf{R}ab}^*(t) = \frac{1}{\sqrt{n_{\mathbf{R}b}^{(0)}(t)(1-n_{\mathbf{R}b}^{(0)}(t))}} \mathrm{Tr}\left(\hat{\Phi}_{\mathbf{R}}^\dagger(\tau)\hat{c}_{\mathbf{R}a}^\dagger \hat{\Phi}_{\mathbf{R}}(\tau)\hat{d}_{\mathbf{R}b}\right) \tag{2.53}$$

Thus again we minimize the action but this time with respect to $|\Psi(t)\rangle$ and $\hat{\Phi}_{\mathbf{R}}(t)$. The fact that also the Slater determinant is to be determined variationally is a very important difference with respect to the equilibrium approximation. Computing the derivatives one obtains the equations of motion for the variational parameters:

$$i\partial_t|\Psi_0(t)\rangle = \mathcal{H}_*[\Phi(t)]|\Psi_0(t)\rangle \tag{2.54}$$

$$i\partial_t\hat{\Phi}_R(t) = \mathcal{H}_R(t)\Phi_{\mathbf{R}}(t) + \langle\Psi_0(t)|\frac{\partial\mathcal{H}_*\left[\hat{\Phi}(t)\right]}{\partial\hat{\Phi}_{\mathbf{R}}^\dagger(t)}|\Psi_0(t)\rangle\rangle. \tag{2.55}$$

These are the time-dependent Gutzwiller equations of motion, which represent a set of coupled differential equations. The equation for the Slater determinant is the that of a non-interacting system with a single-particle Hamiltonian which depends on time, hence the time-dependence of the Slater determinant. The single-particle effective Hamiltonian $H_*\left[\hat{\Phi}(t)\right]$ is indeed a function of the projectors. On the other hand, the time evolution of the projectors is given by a sort of semi-classical expression, where the projectors evolve in time according to the local Hamiltonian and the derivative of the effective single-particle Hamiltonian with respect to the Hermitian conjugate of the field. The two equations realize therefore a mean-field dynamics where the two time-dependent objects feel the presence of the other only through mean-field expectation values. Nonetheless, even this approximate coupling can have important consequences as it couples the low-energy properties with high-energy physcs.

To conclude, we notice that the constraints that hold at $t = 0$ are preserved $\forall t$ by the unitary dynamics. Thus

$$\mathrm{Tr}\left(\hat{\Phi}_{\mathbf{R}}(t)\hat{\Phi}_{\mathbf{R}}(t)\right) = 1 \tag{2.56}$$

$$\mathrm{Tr}\left(\hat{\Phi}_{\mathbf{R}}(t)\hat{\Phi}_{\mathbf{R}}(t)d_{\mathbf{R},a}^\dagger d_{\mathbf{R},a}\right) = \langle\psi(t)|d_{\mathbf{R}a}^\dagger d_{\mathbf{R}a}|\psi(t)\rangle = n_{\mathbf{R}a}^{(0)}(t) \tag{2.57}$$

This is a very important point for practical implementations, because this means that we only need to impose the constraint at the initial time of the dynamics, which can be easily done by solving an equilibrium problem, and the time evolution will preserve it because of its unitarity. On the other hand, a non-unitary dynamics would have required to impose the constraints at any time-step, which is a remarkably harder problem.

2.2.4 Quantum Quench of the Interaction in the Hubbard Model

The time-dependent Gutzwiller approximation formulated above can in principle be applied to any non-equilibrium protocol. In this work we will only present results for quantum quenches of the Hubbard model, but we mention other applications to transport in heterostructures [44], the attractive Hubbard model [32] and a two-orbital model for V_2O_3 [45]. In this section we set up the formalism for the half-filled single-band Hubbard model, while in the next chapter we discuss in details the results obtained with this approach. Also in this case, as we did in equilibrium, we limit to the half-filled case with nearest neighbor hopping, where particle-hole symmetry holds. Thus $\Phi_{i0} = \Phi_{i\uparrow\downarrow} = \Phi_0$, and we assume paramagnetic solutions with $\Phi_{i\uparrow} = \Phi_{i\downarrow} = \Phi_1$. We would like to define $E(t)$ and the effective Hamiltonian $\mathscr{H}^*$ for this case. From (2.53) we obtain

$$R(t) = 2\left[\Phi_0^*(t)\Phi_1(t) + \Phi_1^*(t)\Phi_0(t)\right] \tag{2.58}$$

where the time-dependent amplitudes must be obtained solving (2.57)

$$|\psi(t)\rangle = \exp\left\{-i\bar{\varepsilon}\int_0^t dt'\, R(t')^2\right\}|\text{FS}\rangle \tag{2.59}$$

The Lagrangian is

$$L(t) = 2i\Phi_0^*\dot{\Phi}_0 + 2i\Phi_1^*\dot{\Phi}_1 - E(t) \tag{2.60}$$

with the energy being

$$\begin{aligned} E(t) &= \langle\Psi(t)|P^\dagger H(t)P(t)|\Psi(t)\rangle \\ &= -4\left(\Phi_0^*\Phi_1 + \Phi_0\Phi_1^*\right)^2|\bar{\varepsilon}| + U|\Phi_0|^2 \end{aligned} \tag{2.61}$$

We can rewrite the Gutzwiller parameters as $\Phi_n(t) = \rho_n(t)e^{-i\phi_n(t)}$ and the energy transforms into

$$E(t) = -16\,|\bar{\varepsilon}|\,\rho_0^2\rho_1^2\cos^2(\alpha_0 - \alpha_1) \tag{2.62}$$

A further simplification and a final equation is obtained if we consider the (normalization) relation between ρ_0 and ρ_1

$$2\rho_0^2 + 2\rho_1^2 = 1 \tag{2.63}$$

and $\rho_0^2 = D$ and $\alpha_0 - \alpha_1 = \alpha$. Thus we obtain:

$$E(t) = -16|\bar{\varepsilon}|D\left(\frac{1}{2} - D\right)\cos^2(\alpha) + UD, \tag{2.64}$$

which depends on the two variational parameters α and D. The equations of motions for the two parameters are obtaining requiring stationarity and read

$$\begin{cases} \partial_t \alpha(t) = \frac{1}{2}\frac{\partial E}{\partial D} = -4|\bar{\varepsilon}|\,(1 - 4D(t))\,\cos^2(\alpha) \\ \partial_t D(t) = -\frac{1}{2}\frac{\partial E}{\partial \alpha} = 8|\bar{\varepsilon}|D(1 - 2D)\sin(2\alpha). \end{cases} \tag{2.65}$$

This already shows that, in contrast with the equilibrium Gutzwiller approximation, two parameters are needed to describe the dynamics of the system. This is obviously a slight complication, but at the same time it provides us with a richer picture with respect to equilibrium.

2.3 Dynamical Mean-Field Theory

In this section we make a brief detour from our main trajectory and we give a very short introduction to Dynamical Mean-Field Theory (DMFT), focusing on the aspects which are important for the understanding of the results of the Gutzwiller approximation that we present in the next section. The method and its extension to time-dependent non-equilibrium systems will be presented in another contribution to this book, but we find it useful to discuss some relevant aspects here for consistency. A comprehensive review of DMFT can be found in [19], and more coincise descriptions can be found for example in [46, 47].

Dynamical Mean-Field Theory arises from the pioneering study of the Hubbard model in infinite dimensionality by Metzner and Vollhardt [48], who demonstrated the feasibility of the full perturbation expansion and the momentum-independence of the self-energy. A second crucial step has been the observation by Georges and Kotliar [49] that the method can be seen as a quantum version of a mean-field theory and the lattice model can be mapped onto a self-consistent quantum impurity model.

Nowadays the second point of view is largely dominant. The main idea is that we start from an interacting lattice model with local interactions (the Hubbard model is an obvious example, but, e.g., local electron-phonon coupling as in the Holstein model is an also implemented routinely) and we focus on the local quantum dynamics and thermal behavior. In the mean-field spirit we select a single site as representative of any other, but we retain the full quantum dynamics, as opposed to a static mean-field approach. The quantum evolution of the site is described through the interaction with a non-interacting bath, whose hybridization function is self-consistently determined. Basically the local configuration on the chosen site will fluctuate between all the four local possible configurations with amplitudes which are measured by its Green's function.

In this sense, the only approximation in DMFT is the neglect of spatial fluctuations. As a consequence, the method does not assume anything about the strength of the interactions, and it is completely non-perturbative in all the model parameters. It can be shown that DMFT gives the exact result in any dimensions for the two extreme limits $t = 0$ and $U = 0$.

In the following we give a schematic description of the formalism. We start from the grand canonical partition function of the Hubbard model

$$Z = \int \prod_{i\sigma} Dc^{\dagger}_{i\sigma}\, Dc_{i\sigma} \exp[-S] \tag{2.66}$$

where the action S for the Hubbard model is given by

$$S = \int_0^{\beta} d\tau \left[\sum_{i\sigma} c^{\dagger}_{i\sigma}(\tau)\,(\partial_\tau - \mu)\, c_{i\sigma}(\tau) + \sum_{ij\sigma} t_{ij} c^{\dagger}(\tau) c_{i\sigma}(\tau) + \sum_i U n_{i\uparrow}(\tau) n_{i\downarrow} \right]. \tag{2.67}$$

The idea behind DMFT is to construct an effective theory only for the fermions on a given site that we label as "0". In principle this is realized integrating over all the fermionic degrees of freedom except for those on the chosen site. The integration can be formally carried out and the result is

$$S_{eff} = S_0 + \sum_{n=1}^{\infty} \sum_{i_1 \ldots j_n} \int \eta^{\dagger}_{i_1}(\tau_{i_1}) \ldots \eta^{\dagger}_{i_n}(\tau_{i_n}) \eta_{j_1}(\tau_{j_1}) \ldots \eta_{j_n}(\tau_{j_n}) G^{(0)}_{i_1 \ldots j_n}(\tau_{i_1} \ldots \tau_{i_n}, \tau_{j_1} \ldots \tau_{j_n}) \tag{2.68}$$

$$\stackrel{z\to\infty}{=} \int \eta^{\dagger}_i(\tau_i) \eta_j(\tau_j) G^{(0)}_{ij}(i\omega_n) \tag{2.69}$$

where

$$S_0 = \int_0^{\beta} d\tau \left(\sum_{\sigma} c^{\dagger}_{0\sigma}(\partial_\tau - \mu) c_{0\sigma} + U n_{0\uparrow} n_{0\downarrow} \right) \tag{2.70}$$

and $\eta_i \equiv t_{i0} c_{0\sigma}$. This expression is completely general and formally exact, but it obviously of little use because it includes Green's functions with an arbitrary number of particles, which are basically impossible to compute. DMFT amounts to neglect all the contributions except the first with $n = 1$. In finite dimensions this is simply an approximation, but in the limit of infinite dimensions the $n = 1$ is the only contribution of order 1, while all the higher-order terms vanish as powers of $1/d$.

The result can be written as

$$S_{\text{eff}} = -\int_0^\beta d\tau \int_0^\beta d\tau' \sum_\sigma c^\dagger_{0\sigma}(\tau)\mathcal{G}_0^{-1}(\tau - \tau')c_{0\sigma}(\tau') + U\int_0^\beta d\tau n_{0\uparrow}(\tau)n_{0\downarrow}(\tau), \tag{2.71}$$

where $\mathcal{G}_0$ is a single time-dependent field which describes the effect of the rest of the lattice on site 0. $\mathcal{G}_0$ is usually referred to as the dynamical Weiss field because it generalizes the effective magnetic field of the Weiss mean-field theory of the Ising model.

Just like in the classical mean-field, the effective field can be expressed in terms of the relevant observable, which here is the Green's function of site 0 computed over the effective action

$$G(\tau) = -\langle T_\tau \, c_{0\sigma}(\tau)c^\dagger_{0\sigma}(0)\rangle_{S_{eff}}. \tag{2.72}$$

The self-consistency equation then reads

$$\mathcal{G}_0(i\omega_n)^{-1} = i\omega_n + \mu + G_{loc}(i\omega_n)^{-1} - R[G_{loc}(i\omega_n)] \tag{2.73}$$

where $\omega_n = (2n+1)\pi/\beta$ are the fermionic Matsubara frequencies. $R[G]$ is the inverse function of the Hilbert transform of the non interacting DOS $D(\varepsilon) = \sum_{\mathbf{k}} \delta(\varepsilon - \varepsilon_{\mathbf{k}})$ of the specific lattice considered, namely

$$\tilde{D}(\xi) = \int_{-\infty}^{\infty} d\varepsilon \frac{D(\varepsilon)}{\xi - \varepsilon} \tag{2.74}$$

and $R[\tilde{D}(\xi)] = \xi$.

This is actually the only information about the original lattice within DMFT. This equation clearly shows that we can compute the Weiss field $\mathcal{G}_0$ once we know the Green's function G. On the other hand, we can compute (at least in principle) G from the knowledge of $\mathcal{G}_0$ which defines the effective action. Clearly, while (2.73) is a straightforward calculation, the solution of the effective theory is not equally simple and it represents the bottleneck of the DMFT calculation. In many practical implementations the effective action (2.71) is represented as an impurity model Hamiltonian, in which the site 0 has the same local interaction of the original model and it is hybridized with a non-interacting bath of fermions whose dispersion and hybridization reproduces the Weiss field. The solution of this impurity model requires a numerical "impurity solver", which has to be used iteratively until we find a pair of G and $\mathcal{G}_0$ which satisfy the self-consistency.

We finally consider another equivalent way to write the self-consistency condition which will help us to understand the physical meaning of the DMFT self-consistency.

Starting from (2.73) we can write

$$G(i\omega_n) = R\left[\tilde{D}(G)\right] = \tilde{D}\left[i\omega_n + \mu + G(i\omega_n)^{-1} - \mathcal{G}_0(i\omega_n)^{-1}\right]$$
$$= \int d\varepsilon \frac{D_0(\varepsilon)}{i\omega_n + \mu + G(i\omega_n)^{-1} - \mathcal{G}_0(i\omega_n)^{-1} - \varepsilon}. \tag{2.75}$$

If we define the self-energy of the effective theory as

$$\Sigma(i\omega_n) = \mathcal{G}_0^{-1}(i\omega_n) - G^{-1}(i\omega_n) \tag{2.76}$$

thus

$$G(i\omega_n) = \int d\varepsilon \frac{D_0(\varepsilon)}{i\omega_n + \mu - \Sigma(i\omega_n) - \varepsilon}$$
$$= \sum_k \frac{1}{i\omega_n + \mu - \Sigma(i\omega_n) - \varepsilon_{\mathbf{k}}}, \tag{2.77}$$

which implies that the Green's function of the effective theory has to coincide with the local component (obtained through summing over all momenta) of the lattice Green's function $G(k, i\omega_n)$ computed with the self-energy of the local theory. If compared with the exact solution of the model, we have replaced a momentum dependent self-energy with a momentum-independent one, which can be seen as an average over the full Brillouin zone. As a matter of fact, the locality (momentum-independence) of the self-energy can be taken as an alternative and completely equivalent definition of DMFT.

From the solution of the DMFT equations, we obtain the local Green's function (or any other local observable) and the corresponding Weiss field $\mathcal{G}_0$. Furthermore, the solution provides also the lattice Green's function of the system in the limit of infinite coordination number. It is defined as $G_{ij}(\tau) = -\langle T c_{i\sigma}(\tau) c^{\dagger}_{j,\sigma}(0)\rangle$ or in Fourier space as $G(\mathbf{k}, \imath\omega_n)$. Using the Dyson equation we can write the lattice Green's function in relation to the self energy

$$G(\mathbf{k}, i\omega_n) = \frac{1}{i\omega_n + \mu - \varepsilon_{\mathbf{k}} - \Sigma(\mathbf{k}, i\omega_n)} \tag{2.78}$$

In the limit of infinite coordination number $z \to \infty$, the self energy becomes **k**-independent (in line with mean field procedure) and the only momentum dependence is retained in the dispersion $\varepsilon_{\mathbf{k}}$. Thus, comparing this result with (2.77), we obtain

$$G_{ii}(i\omega_n) = \sum_{\mathbf{k}} G(\mathbf{k}, i\omega_n). \tag{2.79}$$

As a matter of fact DMFT provides a spatial mean-field for the local Green's function, which led to a very powerful description of the Mott transition based on the evolution of the spectral function as a function of the interaction strength. If we start from a non-interacting metallic system with some finite density of states at the Fermi level and a bare bandwidth W, increasing the interaction strength the spectral weight moves towards high energy symmetrically (because of particle-hole symmetry), while the low-energy spectral weight is reduced. Since the momentum-independence of the self-energy implies that the value of the density of states at the Fermi-level is independent on the iteractions, this means that the low-energy feature shrinks as U/t grows, while some spectral features around $\pm U/2$ develop as precursors of the Hubbard bands. The transition occurs when the width of the peak vanishes, leaving an already preformed Mott gap.

The width of the peak is proportional to the quasiparticle weight $Z \equiv \left(1 - \frac{\partial \Sigma''(i\omega_n)}{\partial \omega_n}\right)^{-1}$, which thus vanishes at the Mott transition. Within DMFT the correlation-driven effective mass enhancement $m^*/m = 1/Z$, which means that the Mott transition occurs through a divergence of the effective mass associated with carrier localization. The factor Z plays clearly exactly the same role as the quasi-particle renormalization within the Gutzwiller approximation. As a matter of fact, we can roughly says that the Gutzwiller approximation reasonably reproduces the DMFT picture as far as the metallic peak is concerned, even if it completely fails in reproducing the Hubbard bands.

2.3.1 The Hubbard Model

The paradigmatic model to study correlated electron systems is the celebrated Hubbard model, whose Hamiltonian reads

$$H = -t \sum_{\langle ij \rangle \sigma} \left(c^\dagger_{i\sigma} c_{j\sigma} + h.c. \right) - \mu \sum_{i\sigma} c^\dagger_{i\sigma} c_{i\sigma} + U \sum_i n_{i\uparrow} n_{i\downarrow}, \tag{2.80}$$

a lattice model where $c^\dagger_{i\sigma}$, $c_{i\sigma}$ are creation and annihilation operators for an electron of spin σ on the site i of a given lattice and $n_{i\sigma} = c^\dagger_{i\sigma} c_{i\sigma}$ is the number operator, t is a hopping amplitude which we assume for simplicity to be only between neighboring sites, μ is the chemical potential and U is the interaction between electrons, which is assumed to be purely local. If compared with the electronic structure of actual Mott insulators, the model introduces many simplifications, as it includes one single orbital of s-wave symmetry and it neglects non-local interaction terms and hopping terms beyond nearest-neighboring sites. The choice of a purely local Coulomb interactions is indeed justified only assuming that the long-range Coulomb interaction is screened by the valence electrons and that the relevant orbitals are reasonably localized around the lattice site, which is a natural assumption in a tight-binding description of a solid.

With all these simplifications the physics of the model is controlled basically by two parameters: the ratio between the two energy scales U and t and the number of carriers per site populating the bands $n = \langle 1/N_s \sum_{i\sigma} n_{i\sigma} \rangle$, usually called filling. The former parameter clearly measures the competition between the tendency to form delocalized itinerant states as a consequence of the motion of the electrons and the tendency to localize the charge due to the Coulomb repulsion. In order to better understand this point, we consider the simple situation of one particle per lattice site ($n = 1$), usually referred to as half-filling, because it correspond to a half-filled bands for each spin. We can easily picture the behavior of the sysmte in the to extreme limits $U = 0$ and $t = 0$. For $U = 0$, the hopping term can be diagonalized in momentum space, independently of our choice of nearest-neighbor hopping. In the latter case this gives rise to a cosine dispersion in any spatial direction for a square or cubic lattice. For $n = 1$ the bands are half full, leading to a metal in agreement with the band theory of solids.

In the opposite $t = 0$ limit (atomic limit), the energy is obviously minimized by every real-space configuration where we have one electron in every lattice site. All these states have energy $E = 0$. In this regime the system is trivially an insulator and it remains in an insulating state also for small finite t. Notice that, as long as t is finite, the insulating state is limited to $n = 1$ and for every finite doping $\delta = n - 1$ the pure Coulomb interaction is not able to localize the charge. This gives a clear feeling about the crucial role of the density (or the doping) in determining the properties of the Hubbard model and in general of correlated materials. The crucial role of doping is also confirmed by the experimental phase diagrams of correlated materials, as exemplified by the famous phase diagram of high-temperature superconductors, where superconductivity establishes doping a Mott insulator, and the doping evolution of the system features a variety of anomalous phases.

The simple arguments we gave show that in the two opposite limits the half-filled Hubbard model describes respectively a metal and an insulator. This would imply that a phase transition must take place at some intermediate value of U/t. However, the picture is complicated by the instability towards antiferromagnetic (AFM) ordering at low temperature. Indeed one can easily show that, for the simplest Hubbard model (2.80) with nearest-neighbor hopping only and on a bipartite lattice, the groundstate is always AFM for any finite value of the interaction U. If a more complicated hopping structure or a frustrated lattice are considered, a critical value of the interaction is required to induce AFM ordering. However, this observation does not spoil the expectation of a metal-insulator transition. As we shall discuss in the following, we have convincing evidence that this transition survives at finite temperature in the Hubbard model, which is quite promising in view of application to real materials, where a finite-temperature Mott transition is typically observed.

Despite the formal simplicity, the Hubbard model can not be solved exactly besides the one-dimensional case [27] and in the limit of large coordination, where the Dynamical Mean-Field Theory that we will describe becomes exact. In particular exact solutions for two or three spatial dimensions are not available and the community has resorted to a huge variety of numerical and analytical methods to gain information about the properties of the model. After the proposal by P.W. Anderson that strong correlations in two dimensions are the key to understand the cuprates, this effort has been so strongly intertwined with the hunt for high-temperature superconductivity, to lead to a sort of identification between solution of the problem of high-temperature superconductivity and a solution of the doped Hubbard model.

However, despite a formidable effort, there is no exact solution of the doped two-dimensional Hubbard model and there is no complete consensus about the properties of the model. The reason for this surprising failure is that the model, despite its formal simplicity, features a very profound and archetypal competition between two terms with "opposite" properties. The hopping term leads to delocalized states, which are diagonal in momentum space, while the interaction terms imposes constraints on the solutions based on the real-space configuration and it is already diagonal in real space. For this reason the competition is not trivial and it has defied many attempts of solutions.

However the search for an understanding of high-temperature superconductivity and the realization that strongly correlated materials display also other remarkable properties led to the developement of a variety of techniques designed precisely to deal with strongly correlated electron models, from the Hubbard model to the closely related t-J model. Among these we can list functional renormalization group approaches for two-dimensional correlated systems, a variey of Quantum Monte Carlo methods, the Density Matrix Renormalization Group, the analytical approaches based on slave particles and many others.

A major breakthrough in this direction is represented by the development of the Dynamical Mean Field Theory (DMFT) [19], an approach which attacks the problem by freezing spatial fluctuations but retaining the full local quantum dynamics thus realizing a quantum version of classical mean-field theories. This approach has the great advantage to avoid any assumption about the hierarchy between the energy scales. In this work we will briefly describe the DMFT as a sort of reference method that we will compare with the Gutzwiller approximation, a simple variational approach which however retains many aspects of the DMFT solutions. Last, but not least, these two approaches can be simply generalized out of equilibrium to study the real-time dynamics of correlated materials, which is the main topic of the present manuscript. Before introducing briefly DMFT and the Gutzwiller approximation in more details, we discuss some simple properties of the Hubbard model which will be useful in the following.

2.3.2 Antiferromagnetism in the Hubbard Model

The phenomenology of strongly correlated materials and in particular of Mott insulators suggests some strong link between Mott physics and antiferromagnetic (AFM) ordering of the spins of the localized electrons. An AFM ordering is indeed common to many correlated materials ranging from V_2O_3 to the undoped superconducting cuprates. As a matter of fact, one can show that the groundstate of the half-filled Hubbard model is AFM for any value of the interaction U if the hopping is restricted to nearest neighboring sites. In this section we present some basic calculation and introduce the main concepts about the AFM solution of the Hubbard model and its interplay with the Mott transition that we briefly described in the previous sections.

We start our treatment from the weak-coupling regime $U \ll t$. One can indeed show within linear-response theory, that the AFM susceptibility diverges for any non-zero U. This suggests that a mean-field decoupling which allows for AFM ordering will be energetically stable with respect to a metallic solution and any other ordering. The Hartree-Fock approximation amounts to decouple the quartic interaction term into bilinears as follows:

$$U \sum_i n_{i\uparrow} n_{i\downarrow} \rightarrow U \sum_i \left[\langle n_{i\uparrow} \rangle n_{i\downarrow} + n_{i\uparrow} \langle n_{i\downarrow} \rangle - \langle n_{i\uparrow} \rangle \langle n_{i\downarrow} \rangle \right] \tag{2.81}$$

Antiferromagnetic ordering implies that the mean-field amplitudes at the different lattice sites are related and that they can be expressed in terms of two parameters, the average density per site n and the staggered magnetizations $m_s = 1/N \sum_i (-1)^i$ $(n_{i\uparrow} - n_{i\downarrow})$, which plays the role of the order parameter for the antiferromagnetic state.

$$\langle n_{i\uparrow} \rangle = n + (-1)^{R_i} m_s \tag{2.82}$$
$$\langle n_{i\downarrow} \rangle = n - (-1)^{R_i} m_s \tag{2.83}$$

The mean-field Hamiltonian is now the sum of two Hamiltonians for the up and down fermions, which are coupled because the expectation values of the densities of one spin appear in the Hamiltonian for the other spin. Furthermore the two Hamiltonians are bilinear, hence they can be diagonalized in momentum space. We skip the details of the calculation, which are standard. The main result is that the Brillouin zone is halved because of the two-site unitary cell. In the reduced Brillouin zone we have two bands of dispersion $\pm\sqrt{\varepsilon_k^2 + \Delta^2}$, where $\Delta = U m_s$. At half-filling we populate just the lower band, and we obtain an effective band insulator with a gap of 2Δ. The value of Δ is found minimizing the energy, which leads to the equation

$$\frac{1}{U} = \frac{1}{N_s} \sum_k \frac{1}{\sqrt{\varepsilon_k^2 + (U m_s)^2}}. \tag{2.84}$$

It is easy to prove that, for nearest-neighbor hopping one obtains a finite Δ for any non-zero U, even if Δ is exponentially small in U. This means that, strictly speaking, the groundstate of the Hubbard model is an antiferromagnetic insulator in the limit of small U. However, the insulating behavior is not associated to Mott physics, but it is a Fermi-surface effect due to the nesting of the Fermi surface. This also means that, when we exceed the Néel temperature T_N and the gap closes, we recover a very good, basically uncorrelated metal. These results closely remind the mean-field theory of superconductors. In a few lines we will see that this similarity is far from accidental.

Let us consider the opposite, strong-coupling limit $|U|/t \gg 1$. For a half-filled system the large energetic cost of double occupancy drives the system into a ground state configuration with one electron localized on each lattice site, where the only degree of freedom left is the spin of the localized electrons. If the hopping amplitude is zero, all the configurations with one electron per site are completely degenerate, leading to an extensive entropy.

As soon as a small finite hopping is included, we can use degenerate perturbation theory considering the hopping Hamiltonian as a perturbation of the atomic Hamiltonian. A simple argument can provide us a clear insight about the effect of the perturbation. Let us consider two configurations which only differ in two sites. The first has two up spins in these two sites (second and third explicitly reported in our example), while the second configuration has one up and one down spin in the same sites:

$$|\ldots \uparrow\downarrow\uparrow\downarrow\uparrow\downarrow\ldots\rangle$$
$$|\ldots \uparrow\uparrow\uparrow\downarrow\uparrow\downarrow\ldots\rangle$$

The first-order correction to the energy vanishes because the hopping can not connect two different configurations with one electron per site. The second-order correction to the energy of the i-th groundstate is given by

$$\Delta E_i = \sum_n \frac{\langle i|H_t|n\rangle \langle n|\mathcal{H}_t|i\rangle}{E_0 - E_n}, \tag{2.85}$$

where H_t is the hopping term of the Hamiltonian, $|n\rangle$ are the excited states of the atomic Hamiltonian which are connected to the groundstate via a hopping and E_n are the corresponding eigenvalues. $E_0 = 0$ is the degeneratate groundstate. It should be clear that the first configuration has no contribution from the two sites we selected because Pauli principle forbids any hopping between them. On the other hand, the second configuration can benefit of hoppings to intermediate states in which a doubly occupied site is created on one of the two sites. These states have energy U and the matrix elements in the numerator give t^2. As a result, the second configuration gains an energy $\Delta E = -2t^2/U$ with respect to the first where the factor 2 counts the possible hopping events (up or down spin hopping).

This means that, for every bond, we can gain some energy if the spins are antiparallel. Clearly a state in which all the bonds have antiparallel spins has the lowest energy among all the original degenerate states and it is the groundstate of the Hamiltonian within perturbation theory. Indeed this result can be put un much firmer ground as one can map the half-filled Hubbard model onto a Heisenberg model.

$$\mathcal{H} = \sum_{ij} \frac{4t^2}{U} \left(\mathbf{S}_i \mathbf{S}_j - \frac{1}{4} n_i n_j \right) \tag{2.86}$$

In this case the Néel temperature must be proportional to the exchange constant $\frac{4t^2}{U}$, i.e. it decreases for large $U \gg t$. Assuming that the weak and strong coupling curves for the Néel temperature join smoothly, we immediately understand that the critical temperature for magnetic ordering must reach a maximum for intermediate values of U/t.

However, in the strong-coupling regime, the magnetic ordering is not necessary to make the system insulating. As a matter of fact, we have shown that magnetism

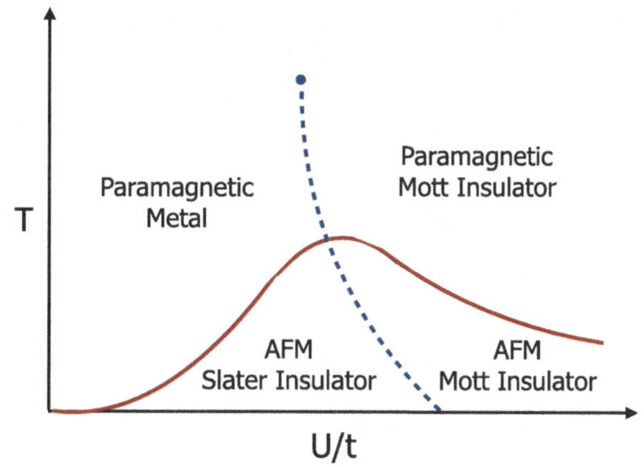

Fig. 2.1 Schematic Phase Diagram of the Repulsive Hubbard model at half-filling. The solid red line marks the Néel temperature for the antiferromagnetic ordering, while the dashed blue line is the Mott transition which ends in a finite-temperature critical point denoted by a blue dot

arises as the ordering of localized spins of the electrons. Therefore the system is now insulating also above T_N and the insulator melts only at very large temperatures of order U. This is indeed the regime of Mott physics, where the insulating behavior is driven by charge localization and the magnetic ordering occurs only at much lower temperatures due to the effect of the effective magnetic constants.

Therefore, despite the low-temperature phase diagram is dominated by the antiferromagnetic ordering, the Mott transition is still expected to occur at finite temperature separating a metallic state for low U/t and an insulating paramagnetic state for large U/t. In Fig. 2.1 we report a schematic phase diagram in a U-T phase diagram, where the generic diagram for the antiferromagnetic ordering is plotted together with the Mott transition line as it is predicted by Dynamical Mean-Field Theory and Gutzwiller approximation.

It must be noted that the absolute values of T_N strongly depend on the specific lattice. In actual systems one has always a certain degree of frustration which leads to a reduction of T_N and to a diagram qualitatively similar to our Fig. 2.1, where the Mott transition line emerges from the antiferromagnetic dome, leading to a sizable window of temperatures where a direct transition between a paramagnetic metal and a Mott insulator takes place without any role of magnetic ordering. At temperatures higher than the critical point we mark with a blue dot the system evolves smoothly from weak to strong coupling and only a crossover connects solutions with a more metallic and a more insulating character.

2.3.3 The Attractive Hubbard Model

In this section we perform a little detour, which will prove very helpful. We will indeed show that one can map a repulsive Hubbard model onto an attractive Hubbard model, i.e., a model which is formally identical to the Hubbard model where the on-site repulsion is replaced by an on-site attraction. As we will see the two models map exactly onto each other at half-filling when the magnetization is zero, and the mapping exchanges the roles of spin and charge degrees of freedom. We start from out usual Hubbard model, which we consider in a grandcanonical ensemble with a chemical potential μ and a magnetic field h. The reason to introduce explicitly the latter term will become clear in the following.

$$H = -t \sum_{\langle i,j \rangle} c^{\dagger}_{i\sigma} c_{j\sigma} + U \sum_{i} n_{i\uparrow} n_{i\downarrow} - \left(\mu' + \frac{U}{2}\right) \sum_{i} (n_{i\uparrow} + n_{i\downarrow}) - h \sum_{i} (n_{i\uparrow} - n_{i\downarrow}) \tag{2.87}$$

where $U > 0$ and h is the magnetic field. All the other quantities have been defined before. We assume that the lattice is bipartite, i.e., it can be divided in two sublattices A and B such that the hopping only connects sites of sublattice A with sites of sublattice B. This is obviously the case of square and cubic lattices and of the body-centered-cubic. We introduce the following unitary transformation

$$c_{i\uparrow} \rightarrow c_{i\uparrow} \tag{2.88}$$

$$c_{i\downarrow} \rightarrow (-1)^{R_i} c^{\dagger}_{i\downarrow} \tag{2.89}$$

where $(-1)^{R_i}$ is essentially a sign which assumes one value on one sublattice and the opposite on the other.

It is easy to see that the hopping term remains unchanged under the transformation. This is trivial for the hoppings of the up spins, while the equality holds for down spins because of the staggered signs $(-1)^{R_i}$. Indeed the mapping transforms any $c^{\dagger}_{i\downarrow} c_{j\downarrow}$ into $(-1)^{(R_i+R_j)} \tilde{c}_{i\downarrow} \tilde{c}^{\dagger}_{j\downarrow}$. Since the two sites connected by the hopping belong by construction to two different sublattices, the prefactor is -1 and it cancels the sign associated to the fermionic permutation that reestabilishes the standard order of fermionic operators. Therefore we obtain a term identical to the original hopping only with site indices swapped. Since all the pairs of sites are included, this demonstrates that the hopping is unchanged.

As far as the density on a given site is concerned, we have $n_{i\downarrow} \rightarrow 1 - \tilde{n}_{i\downarrow}$. This means that the total local density now maps into the the local magnetization along the z-axis (-1) and the magnetization maps onto the density $(+1)$

Therefore the Hubbard Hamiltonian transforms into (we now drop the tilde to simplify the notations).

$$H = -t \sum_{\langle i,j \rangle} c^{\dagger}_{i\sigma} - U \sum_{i} n_{i\uparrow} n_{i\downarrow} - \left(h + \frac{U}{2}\right) \sum_{i} (n_{i\uparrow} + n_{i\downarrow}) - \mu' \sum_{i} (n_{i\uparrow} - n_{i\downarrow}) \tag{2.90}$$

The crucial effects of this transformation are that the interaction term has changed sign $U \rightarrow -U$, whereas the role of the chemical potential and the magnetic field are interchanged. More precisely the deviation from half-filling (doping) maps onto the magnetization and so do the corresponding fields μ' and h.

In particular if $\mu' = h = 0$, which corresponds to half-filling and zero magnetization, also the attractive model is in the same conditions. Therefore the half-filied non magnetic repulsive Hubbard model maps onto the attractive Hubbard model in the same conditions. A doped repulsive repulsive Hubbard model maps instead on a magnetized attractive Hubbard model.

The mapping implies that all the informations we obtain for one of the two models can be used to extract information about the other.

For example, we have shown that the groundstate of the repulsive model is antiferromagnetic for every value of U/t All the expectation of the repulsive model, as well as the order parameter, shall be translated in the attractive one. The three components of the staggered magnetization become

$$\langle m_s^z \rangle = \frac{1}{N_s} \sum_i (-1)^{R_i} \langle n_{i\uparrow} - n_{i\downarrow} \rangle \rightarrow \frac{1}{N_s} \sum_i (-1)^{R_i} \langle n_i \uparrow + n_{i\downarrow} - 1 \rangle \quad (2.91)$$

$$\langle m_s^x \rangle = \frac{1}{N_s} \sum_i (-1)^{R_i} \langle c_{i\uparrow}^\dagger c_{i\downarrow} + c_{i\downarrow}^\dagger c_{i\uparrow} \rangle \rightarrow \frac{1}{N_s} \sum_i \langle c_{i\uparrow}^\dagger c_{i\downarrow}^\dagger + c_{i\downarrow} c_{i\uparrow} \rangle \quad (2.92)$$

$$\langle m_s^y \rangle = \frac{1}{N_s} \sum_i (-1)^{R_i} i \langle c_{i\uparrow}^\dagger c_{i\downarrow} - c_{i\downarrow}^\dagger c_{i\uparrow} \rangle \rightarrow \frac{1}{N_s} \sum_i i \langle c_{i\uparrow}^\dagger c_{i\downarrow}^\dagger - c_{i\downarrow} c_{i\uparrow} \rangle \quad (2.93)$$

Clearly we can recombine the second and third term obtaining the real and imaginary part of the complex order parameter

$$\Delta = \frac{1}{N_s} \sum_i \langle c_{i\uparrow}^\dagger c_{i\downarrow}^\dagger \rangle, \quad (2.94)$$

which is the s-wave superconducting order parameter. Hence the three components of the staggered magnetization transform into a staggered charge-density wave described by (2.91) and an s-wave order parameter. Both states are indeed naturally expected in an attractive Hubbard model, which can favor either formation of Cooper pairs or the formation of doubly occupied sites and their staggered ordering in space.

The mapping between the two model implies that the half-filled non magnetic attractive Hubbard model has a three-fold degenerate groundstate where the system can be in any linear combination of the CDW and superconducting groundstates. The mapping we discussed allows us also to understand what happens doping the attractive Hubbard model away from half-filling. The doping obviously corresponds to a shift of the chemical potential with respect to the particle-hole symmetric point for the attractive model. We can use the mapping "backwards" to gain insight on the corresponding physics. The chemical potential change of the attractive model becomes in fact a magnetic field along the z direction of the spin for the repulsive model. This field obviously competes with the antiferromagnetic alignment of the spins along the z direction, while the antiferromagnetic ordering along the planes is basically unaffected. This means that the repulsive model is expected to display a planar AFM ordering. Turning back to the attractive model, this implies that the

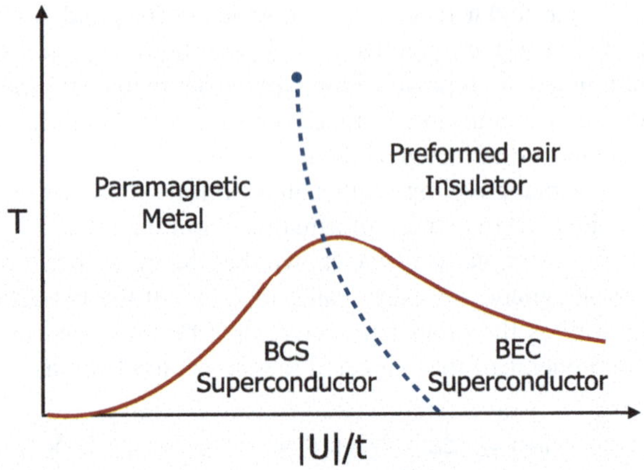

Fig. 2.2 Schematic Phase Diagram of the attractive Hubbard model (at any filling). The solid red line marks the critical temperature for s-wave superconductivity, while the dashed blue line is a pairing transition line separating a metal from an insulating state of preformed pairs

degeneracy between the charge-density wave and the s-wave superconductor is broken in favour of the latter. This means that s-wave superconductivity is stable at any filling in the attractive Hubbard model and it is only degenerate with commensurate charge ordering at half-filling. Moreover, we can also map the Mott transition of the repulsive Hubbard model into the attractive language. The Mott insulator is a state where singly occupied sites are overwhelming over empty and doubly occupied. In the attractive model, this becomes a state where doubly occupied and empty sites dominate and basically all the electrons are paired-up. We can view this as a phase of incoherent "preformed" pairs which fail in becoming superconducting becase they lack phase coherence [28]. The Mott transition then becomes a "pairing transition" connecting a metal with such paired state [29–31]. Finally, within the superconducting phase, we have a crossover between a BCS superconductor at weak coupling and a Bose-Einstein condensation (BEC) regime at strong coupling, which is the attractive counterpart of the evolution between a Slater antiferromagnetism described within Hartree-Fock and the Heisenberg magnetism at strong coupling. The resulting phase diagram is reported in Fig. 2.2.

In the context of this manuscript, this analysis shows that all the information we can gain on the repulsive Hubbard model and its non-equilibrium dynamics can be used to study also the attractive Hubbard model and viceversa. As far as this manuscript is concerned, we will present results for the repulsive model which overwhelmingly more numerous than those for the attractive model. The mapping implies their direct relevance also for the attracrive model. We mention for completeness a recent Gutzwiller study [32] and a DMFT study of the attractive Hubbard model in an electric field [33].

2.4 Quantum Quenches for the Hubbard Model

In this section we discuss how the time-dependent Gutzwiller approximation describes the non-equilibrium physics of strongly correlated systems. We focus on the single-band Hubbard model that we introduced in the previous sections and we compare the results with those of time-dependent Dynamical Mean-Field Theory.

Out of Equilibrium Protocols

In Fig. 2.3 we report the schematic phase diagram of the Hubbard model at half-filling, which is characterized by a Mott transition in the absence of magnetic ordering and a low-temperature "bell-like" antiferromagnetic critical temperature. Let us assume to perturb the system starting from the non-interacting limit and to switch on the interaction up to some final value. A completely different outcome can be expected according to the protocol we use to swich on the interaction. If the interaction is increased very slowly (i.e. adiabatically), the system will slowly evolve through its equilibrium phase diagram, becoming first a correlated metal and then a Mott insulator when the final value of the interaction crosses the equilibrium critical U. The situation is much more open if the interaction is switched on rapidly on some typical timescale τ^*, or even abruptly, as in a quantum quench. In these cases there is no reason why the system should move through its equilibrium phase diagram and novel phenomena can be expected. For relatively long ramping times τ^* we can expect that the system will reach an asymptotic thermal state with an effective temperature $T*$ which grows as the ramping time is reduced, i.e. the interaction is switched-on more abruptly. This means that instead of moving along a horizontal line in the U-T diagram, the system will follow a trajectory where also the temperature changes, as depicted as a green line in the figure. Because of the shape of the Mott transition line (dashed blue line), this trajectory will then cross the equilibrium transition line at a smaller value of U with respect to the equilibrium transition. Therefore we can expect that a dynamical phase transition will occur at some $U^*(\tau^*)$ which decreases as the ramping time decreases.

It is not equally simple to predict the behavior of the system in the limit $\tau^* \to 0$, the quantum quench, where it is not even clear if we can understand the physics in terms of our knowledge of the equilibrium properties of the model. In the following we present the results obtained within the Gutzwiller approximation, which we will then compare with the DMFT results.

2.4.1 *Quench Dynamics in the Gutzwiller Approximation*

We start from the equations for the quantum quench from $U = 0$ to a finite U that we derived in Sect. 2.2.4. Solving (2.65) we can obtain the dynamics of the double occupancy $D(t)$ and of the quasiparticle weight $Z(t)$, which we report in Fig. 2.4

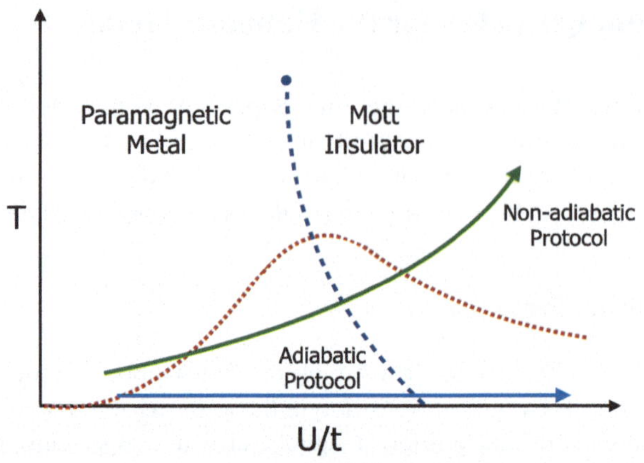

Fig. 2.3 Schematic diagram to illustrate how non equilibrium protocols can be connected with the equilibrium phase diagram of the Hubbard model. The blue line depicts the way an adiabatic switching of the interaction allows to move following a fixed-temperature line. The green line mimicks a non-adiabatic protocol where the excitation also enhances the effective temperature. The red dotted line is the Néel temperature and the blue dashed line marks the Mott transition

for four values of U, which we refer to the critical value for the equilibrium Mott transition U_c.

For values of the quench different from $U_c^{dyn} = U_c/2$, $D(t)$ and $Z(t)$ display an undamped oscillatory behaviour, while at $U^* = U_c/2$ the dynamics shows a relaxation towards a stationary solution of the double occupation and consequently of the quasiparticle weight. The period of the oscillations increases when we approach $U*$ from both directions. In particular we can prove that it diverges logarithmically at U_c^{dyn} from both above and below the critical value: $\tau \sim 1/\log|u_c^{dyn} - u_f|$, with $u = U/U_c$, while in the limit of very large quenches it approaches the atomic limit $\tau = 2\pi/U_f$.

These results clearly show the existence of some kind of dynamical phase transition characterized by a divergent period of oscillation at $U* = U_c/2$. However, the fact that both phases display undamped oscillations is likely a shortcoming of the mean-field character of the Gutzwiller approximation, which lacks dissipation terms which are crucial to lead to thermalization or anyway to a stationary or quasi-stationary behavior.

However, despite this limitations, the Gutzwiller approximation contains an information about the different phases, which can be extracted from the long-time average of a relevant observable O, defined as

$$\langle O \rangle_t = \lim_{\tau \to \infty} \frac{1}{\tau} \int_0^\tau dt\, O(t) \tag{2.95}$$

The results are shown in Fig. 2.5, again for Z and D. It is evident that, despite the oscillations, the long-time averages both vanish at the transition point, while they are finite for any other value. This implies that the strong-coupling side of the transition is not simply the equilibrium Mott insulator, which would have a vanishing (or small because of the finite effective temperature) for Z and a small value of D. At $U/U_c = 1/2$ the two quantities diverge as

$$\langle D\rangle_t(u_f), \langle Z\rangle_t(u_f) \sim 1/\log\left(|u_c^{dyn} - u_f|\right), \tag{2.96}$$

which confirms that U_c^{dyn} can be considered a dynamical critical point because it separates two dynamical phases. To some extent we can still consider it as a dynamical counterpart of the equilibrium Mott transition despite some important differences.

For small quenches we find analytically the following expressions which we can compare with the equilibrium values $D_{eq}(u_f)$ and $Z_{eq}(u_f)$

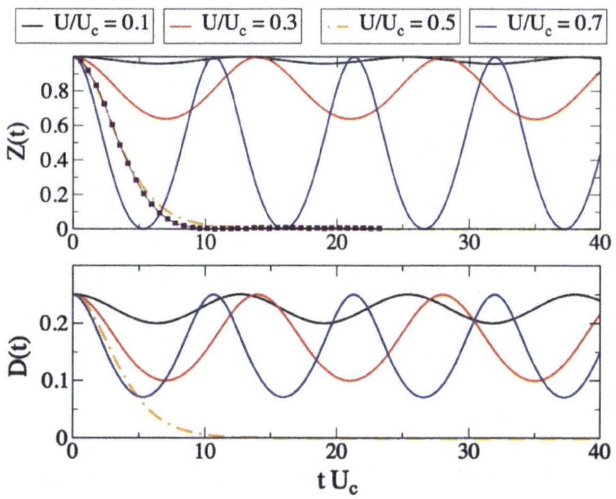

Fig. 2.4 Dynamics of the quasiparticle weight $Z(t)$ (upper panel) and the double occupation $D(t)$ (lower panel) for different values of the quenched parameter U. The dotted line in the upper panel represents the result obtained from DMFT calculation of [50] in proximity of the critical value (from [40])

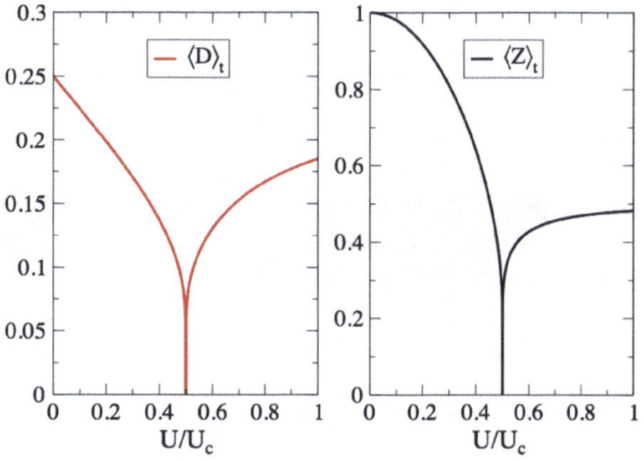

Fig. 2.5 Long-time average of the double occupation $\langle D\rangle_t$ (left panel) and the quasiparticle weight $\langle Z\rangle_t$ (right panel) as defined in the main text as functions of the final value of the quenched interaction U. At the dynamical transition point $U/U_c = 0.5$ both the quantities vanish with a logarithmically divergent slope (from [40])

$$\langle D\rangle_t(u_f) = \frac{1-u_f}{4} = D_{eq}(u_f) \tag{2.97}$$

$$\langle Z\rangle_t(u_f) = 2Z_{eq}(u_f) - 1, \tag{2.98}$$

which means that, as far as the number of doubly occupied sites is concerned, the system approaches the equilibrium value at zero temperature, but the quasiparticle renormalization is clearly different. Therefore, despite the behavior of the double occupancy, the system did not reach the equilibrium state, but it is trapped in a metastable state, where momentum-dependent quantities such as $Z_{eq}(u_f)$ have not thermalized, while momentum-integrated quantities like $D_{eq}(u_f)$ did. The lack of full thermalization is not surprising in view of the limitations of the Gutzwiller dynamics.

These results, where the system is trapped in a state which can not be obtained in equilibrium are in agreement with the prethermalization regime obtained through perturbative calculations [51, 52] for weak quenches. The main motivation adduced for the existence of a prethermal regime in the limit of weak quenches is related to the integrable-like feature of the Hubbard Hamiltonian in the non-interacting limit. Further considerations for the effective temperatures that can be defined to construct a Generalized Gibbs Ensemble has been performed in [53].

In order to shed light on the physics of the phases separated by the dynamical transition we report the evolution of the spectral functions $A(\omega, t)$, which can help us to understand to which extent the two phases separated by the quantum phase transition are similar to the equilibrium metal and Mott insulators [54]. The spectral function can be computed as

$$A(\omega, t) = -\frac{1}{\pi}\mathrm{Im}\int ds\, e^{i\omega s}\, G^{\mathbf{R}}(t+s, t) \tag{2.99}$$

where we introduced the two-time local retarded Green's function

$$G^{\mathbf{R}}(t+s, t) = -\theta(s)\langle\Psi(0)|\{c_{\mathbf{R}}(t+s), c^{\dagger}_{\mathbf{R}}(t)\}|\Psi(0)\rangle \tag{2.100}$$

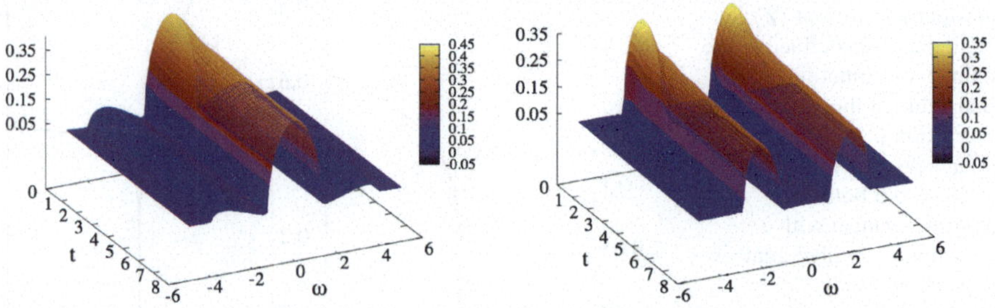

Fig. 2.6 Spectral functions $\bar{A}(\omega, t)$ as a function of time and the frequency for $U/U_c = 0.4$ (below the dynamical transition) in Fig. 1.4.1 and $U/U_c = 0.7$ (above the dynamical transition) in Fig. 1.4.1 (from [40])

The result obtained for these quantities using the Gutzwiller approach is

$$G^{\mathbf{R}}(t+s,t) = -\theta(s)R(t)R(t+s)\int d(\varepsilon)\,\rho\varepsilon\, e^{-i\int_t^{t+s} d\tau\,\varepsilon\, Z(\tau)} \tag{2.101}$$

where the original Fermi operator $c_{\mathbf{R}}$ is replaced by the renormalized fermion, i.e., multiplied by the time-dependent renormalization factor Z.

Once again, since the Gutzwiller approximation can not capture the relaxation behaviour, the long-time average of the spectral function $\bar{A}(\omega,t) = 1/t\int_0^t A(\omega,\tau)$ is considered. Figure 2.6 shows the behaviour of the spectral function according to the selected value of the interaction above and below the dynamical critical point U_c^{dyn}. As shown in the Figure, below the dynamical transition a well defined peak is centered at $\omega = 0$, thus indicating a coherent metallic nature of the state. Instead, for values of the final quench above the dynamical transition, two well-separated Hubbard bands are clearly present, which lie approximately at $\pm U/2$. This demonstrates that the picture that the dynamical transition separates two different dynamical regimes that can be interpreted as the non-equilibrium counterpart of the Brinkmann-Rice transition.

We have thus shown that the "mean-field-like" Gutzwiller approximation provides a consistent picture for a dynamical phase transition despite the lack of thermalization descending from the unitary character of the time evolution. As a matter of fact, the observables keep on oscillating also at very long times, but their long-time averages have two different dynamical behaviors on the two sides of a dynamical phase transitions where these averages vanish. We now compare with DMFT, where the time evolution is less restricted.

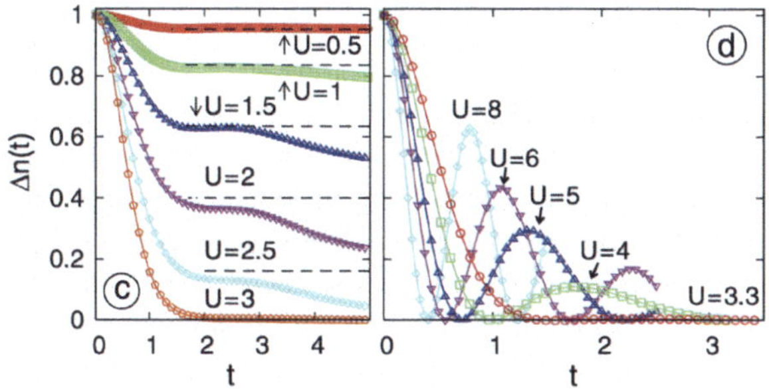

Fig. 2.7 Dynamical Mean-Field Theory results for an interaction quench in the half-filled Hubbard model. We report the time dependence of the non-equilibrium contribution to the momentum distribution at the Femi surface for different values of the final interaction U. Panel (**c**) contains results for small U where prethermalization occurs, while panel (**d**) reports large-U results. The dashed lines in panel (**c**) mark the prethermal plateau of (2.98) (from [50])

2.4.2 Dynamical Phase Transition Within DMFT

The dynamical phase transition for the half-filled Hubbard model has been indeed first discovered and studied using non-equilibrium DMFT calculations by Eckstein and Werner [50]. The main results are shown in Fig. 2.7 where the quantity of interest is the jump of the Fermi distribution at the Fermi level, which, within a Fermi liquid framework, is the analogous of $Z(t)$ considered in the previous Gutzwiller analysis.

Two completely different behaviors are apparent: for small quenches (left panel) there are no oscillations and the dynamics presents a first decrease followed by a plateau which can be associated to a prethermalization dynamcs, whose evolution is in agreement with (2.98) as shown by the dashed lines in the figure. Finally at longer times the systems exits the prethermal region and approaches a long-time limit. For large quenches the dynamics is dominated by oscillations with period $2\pi/U$ superimposed to a clear damped dynamics.

Since time-dependent DMFT requires a numerical solution, in this case it is not possible to pinpoint the critical point analytically as in the Gutzwiller approximation, but the existence of two completely different dynamics is undisputable. Also in this case, the dynamical phase transition reminds of the equilibrium Mott transition in the sense that the two dynamical regimes are clearly connected with metallic and Mott insulating solutions, but the connection is not so obvious. For example, if we estimate the effective temperature T^* based on the energy injected in the system via the quench, we resulting T^* is much larger than the critical temperature for the equilibrium Mott transition, which would not justify a critical behavior.

2.4.3 From Adiabatic Switching to Quantum Quenches: Ramping up the Interaction

In this section we explore the evolution from an adiabatic switch of the interaction, where the system is bound to follow the equilibrium phase diagram, to the quantum quench dynamics, where we have seen that intrinsically non-equilibrium processes take place. We can follow this sort of crossover by considering a *ramp protocol* with a characteristic time τ which allows us to tune the degree of adiabaticity of the process. Therefore we consider a Hubbard model with a time-dependent $U(t)$ which evolves according to

$$U(t) = U_i + \frac{t\Delta U}{\tau} \quad 0 \leq t \leq \tau \tag{2.102}$$

$$U(t) = U_f \qquad t \geq \tau \tag{2.103}$$

Clearly we have introduced a furher timescale besides the intrinsic timescales of the correlated system and it leads to a variety of regimes in time. We can split the time

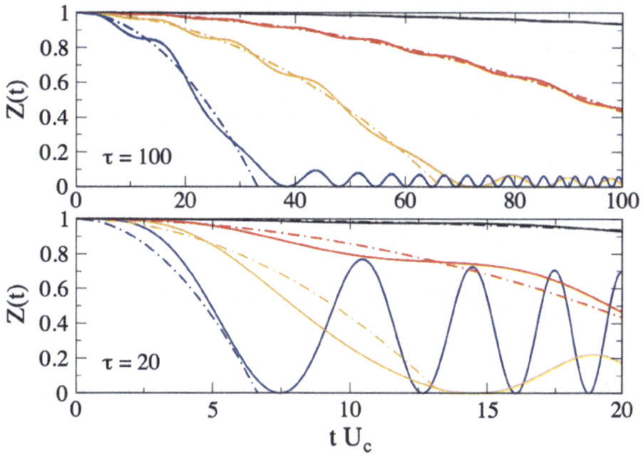

Fig. 2.8 Time evolution of the quasiparticle weight $Z(t)$ obtained within the time-dependent Gutzwiller approximation for different values of the final interaction $u_f = 0.25, 0.75, 1.5, 3.0$ (from top to bottom) respectively for two different ramp time $\tau = 100$ and 20. U_c is the critical value for the equilibrium Mott transition. The dashed lines represents the adiabatic dynamics Plot from [55]

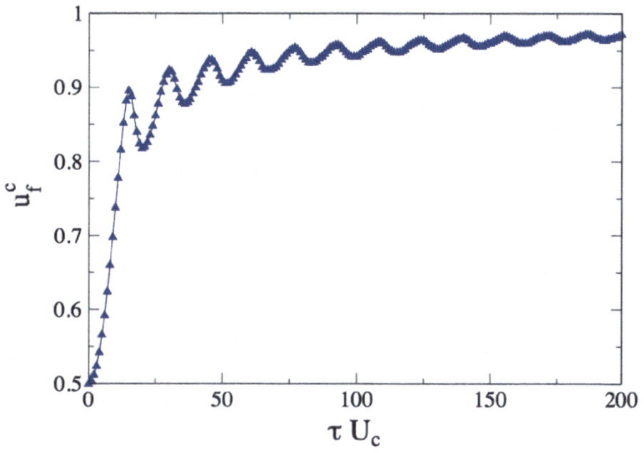

Fig. 2.9 Critical point for the dynamical phase transition u^c_f as a function of the ramp duration τ within the Gutzwiller approximation plot from [55]

evolution in three phases: (i) the driven dynamics during the ramp time $0 < t < \tau$, (ii) the state of the system after the ramp is over and the interaction reached its final value and (iii) the non equilibrium dynamics after the ramp. We refer to previous literature [55] for a detailed analysis of the non-equilibrium processes during the ramping time, while here we focus mainly on the effect of a finite ramp on the location of the dynamical phase transition.

In Fig. 2.8 we report the Gutzwiller mean field dynamics at half-filling of the quasi-particle weight $Z(t)$ i for different values of the final interaction and compared with its adiabatic dynamics obtained assuming that the system stays in its instantaneous variational ground state $Z_{ad}(t) = 1 - u^2(t)$ (Fig. 2.9).

In the two panels we show two different ramping times $\tau = 20$ and 100 (in units of the inverse of the nearest-neighbor hopping). In each plot we consider four values of the interaction. For the larger value of τ the dynamics of the ramped case follows the adiabatic result closely. Deviations only appear for the largest value of U that we consider, $U = 3U_c$, where some deviations and long-time oscillations appear, while the adiabatic evolution reached a Mott insulator with $Z(t) = 0$. Quite naturally, deviations with respect to the adiabatic dynamics appear when we reduce τ.

We now briefly focus on the dependence of the critical interaction strenght for the dynamical transition as a function of the ramping time τ for quenches starting from a non-interacting system. For $\tau \to 0$ we approach the analytical result for the quantum quench $u_c^f = 1/2$. On the other hand, when the ramping time goes to infinity we correctly recover the equilibrium zero-temperature critical interaction, which corresponds to $u_c^f = 1$ as expected for an adiabatic evolution. The evolution between the two regimes mostly occurs in a region of small times and it is accompanied by some fluctuations which are possibly an artifact of the unitary nature of the Gutzwiller time evolution.

2.4.4 *Interaction Quench in the Antiferromagnetic Hubbard Model*

We finally extend our analysis of quantum quenches to the antiferromagnetic solution of the Hubbard model which, as we discussed in Sect. 2.2, is the low-temperature state of the system for any value of the interaction (in the case of nearest-neigbor hopping). Using the equilibrium phase diagram as a reference, we understand that interesting behavior can be expected because of the nonmonotonic behavior of the ordering temperature. Even considering an adiabatic evolution, if we start from an intemediate coupling strength in the middle of the magnetic dome, if the temperature is finite and relatively small, we can melt the magnetic ordering both if we increase the coupling and if we reduce it. We therefore consider and compare two different cases:

$U_f < U_i$: from intermediate to small U

In order to monitor the time evolution of the magnetic phase after the quantum quench we follow the time evolution of the staggered order parameter $m = 1/N_s \sum_i (-1)^{R_i} (n_{i\uparrow} - n_{i\downarrow})$.

In Fig. 2.10 we report the results starting from $U_i = 4.0$, which lies in the middle of the magnetic dome, to final values $U_f < U_i$ ranging from 1.6 to 3.8. In all cases the order parameter drops, but a qualitative difference is evident. For small value of the final interaction $U_f = 1.8, 1.6$ the magnetization rapidly vanishes while for higher U_f the typical behaviour shows a fast decay at short times followed by an oscillatory behaviour. The oscillations happen around a value which is larger than the thermal one corresponding to the final value of the interaction and to the energy that the quench injects into the system. This means that the the dynamics stays trapped in a non-thermal ordered state. We also notice that the asymptotic value is larger than the equilibrium one even when the effective temperature is larger than the Néel temperature.

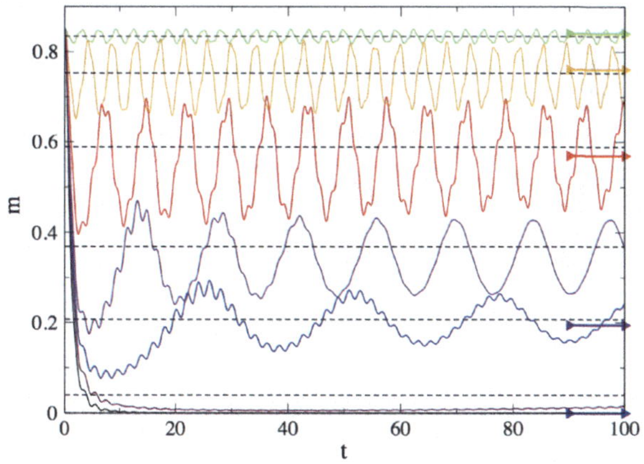

Fig. 2.10 Time evolution of the AFM order parameter for quenches from $U_i = 4.0$ to $U_f = 3.8, 3.2, 2.6, 2.2, 2.0, 1.8, 1.6$ (from top to bottom) obtained within the Gutzwiller approximation. The arrows indicate the thermal values of m corresponding to the final interaction and to the energy pumped into the system by the quench, while the black dashed lines indicate the long-time averages figure from [57]

This results in indeed in agreement with DMFT result for the same protocol, where the resilience of the magnetization in a non-thermal state can be interpreted as a pre-thermalization plateau [56].

$U_f > U_i$: from intermediate to large U

The evolution of the magnetization in the case where the quench ends with a larger magnetization with respect to the starting value.

Indeed in Fig. 2.11 we see that for $U_f = 12$ we find a result analogous to the one we just commented for small final U's, where the magnetization survives in a non-thermal state despite the effective temperature exceeds the equilibrium ordering temperature. Increasing U one observes a clear enhancement of the oscillation period and a reduction of the asymptotic value. Also in this case, when the quench brings the system very far from the starting interaction and in a region of very small equilibrium Néel temperature, one finds a very fast damping of the order parameter to zero.

2.5 Non-equilibrium Superconductors: Mean-Field Theory

The understanding of the behavior of superconductors and superfluids driven out of equilibrium is another extremely active field which encompasses at least two different worlds: Ultracold gases of cold atoms with attractive interactions and pump probe experiments on high-temperature and conventional superconductors. Both fields are very active and in rapid development. In the former case, the system can be thought as an ideal s-wave superconductor where the paring comes from an explicit attractive

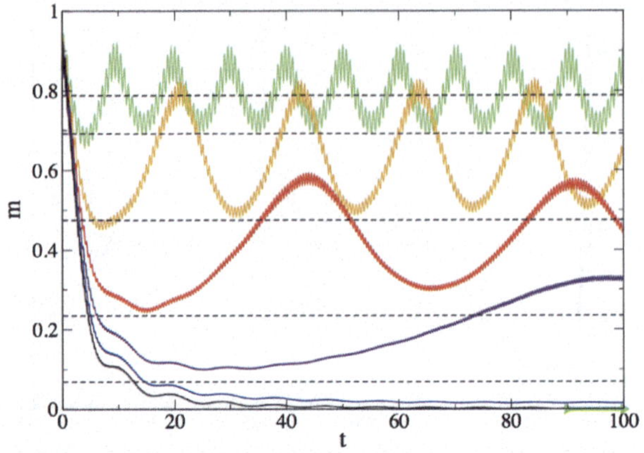

Fig. 2.11 Time evolution of the staggered magnetization m for quenches $U_i = 4.0$ to $U_f =$ 12.0, 14.0, 16.0, 18.0, 20.0, 22.0 (top to bottom). The green arrow indicates the thermal value for $U_f = 12.0$ and shows that the effective temperature has already crossed the equilibrium ordering temperature. The black dashed lines indicate the values of the long-time average figure from [57]

interatomic interaction (tuned by means of a Feshbach resonance) and it can be driven out of equilibrium via a sudden change of a model parameter, which realizes the quantum quench protocol that we introduced in relation to the Hubbard model. On the other hand, in high-temperature superconductors there is still a lively debate on the origin of the superconducting pairing, but the overall consensus is that it derives from strong electron-electron interactions, and that it has a d-wave symmetry, which implies the existence of nodes in the Brillouin zone where the superconducting gap vanishes and quasiparticle excitations are still possible even in the superconducting state. Moreover, the non-equilibrium protocol is not a quantum quench, but it is given by the interaction of the solid with a laser pulse (the pump). The evolution of the system is than probed by a second laser after a given delay time.

Motivated by the two sets of experimental evidences, in this section we present some basic calculations in which we compare the response of s-wave and d-wave superconductors after a quantum quench of the interaction strength. Furthermore we will limit the treatment to weak-coupling, where we can use the simple BCS approximation. For the reasons we mentioned above, the model and protocol that we study are directly relevant to cold-atoms experiments, where the interaction can be tuned to be weak and attractive, and it can be suddenly switched on as in a quantum quench dynamics. The relevance to high-Tc superconductors in instead much less direct, because our approach neglects strong correlation effects and the possible strong-coupling nature of the pairing together with effects like the pseudogap, Mott physics, spin fluctuations and deviations from Fermi-liquid. Moreover, a quantum quench is only a crude modeling of the actual non-equilibrium protocol realized in pump-probe spectroscopies. However, the comparison between s-wave and d-wave superconductors will allow us to single out the role of the existence of nodal regions in the complex response of high-Tc supercoductors. This chapter is based on [58, 59] and references therein.

For this reason the main focus of this section will be on the symmetry of the order parameter, and we will limit our study to a standard mean-field treatment of a superconductor. In the s-wave case we may consider the mean-field model as either arising from an electron-boson interaction, a role which is played by the phonons in the BCS theory, or a mean-field treatment of the attractive Hubbard model we introduced previously. In this sense, the quantum quench of a s-wave superconductor is indeed related to the quench of an antiferromagnet that we discussed previously within the Gutzwiller approximation, even if it must be kept in mind that the superconducting order is described by a two-dimensional order parameter (a complex number), while the antiferromagnetic order parameter with a magnetization along the z axis is a one-dimensional real number. As far as the d-wave model is concerned, we may derive it via a mean-field decoupling of the t-J model, one of the most popular (and simplest) models to describe copper-based high-temperature superconductors.

Our results will be relevant for the so-called BCS Hamiltonian

$$H_{\text{BCS}} = \sum_{k\sigma} \varepsilon_k c^\dagger_{k\sigma} c_{k\sigma} - \lambda \sum_{kp} \gamma_k \gamma^*_p c^\dagger_{k\uparrow} c^\dagger_{-k\downarrow} c_{-p\downarrow} c_{p\uparrow}, \tag{2.104}$$

where $c^\dagger_{k\sigma}$ is the creation operator of electrons with momentum k, spin projection σ. ε_k is the non-interacting dispersion, and the strenght of the attractive interaction is λ, which is a positive number with our notation. The Hamiltonian features an attractive interaction which is separable in momentum space. This means that the interaction term only depends on two momenta, as opposed to the three independent momenta of a generic interaction, and that the dependence on the two remaining momenta is factorized in the product of two identical functions with the same momentum dependence γ_k, which defines the momentum structure of the interaction and consequently of the pairing gap.

For a large number of particles, this Hamiltonian can be shown to be equivalent to a mean-field decoupled Hamiltonian

$$H_{\text{MF}} = \sum_{k\sigma} \varepsilon_k c^\dagger_{k\sigma} c_{k\sigma}, - \sum_k (\gamma_k \text{d} c^\dagger_{k\uparrow} c^\dagger_{-k\downarrow} + \gamma^*_k \text{d}^* c_{-k\downarrow} c_{k\uparrow}) \tag{2.105}$$

where the two-particle interactions are replaced by single-particle terms with mean-field amplitudes

$$\text{d} = \lambda \sum_p \gamma^*_p \langle c_{-p\downarrow} c_{p\uparrow} \rangle, \tag{2.106}$$

which is nothing but the superconducting order parameter associated with pairs with symmetry given by γ_k.

The structure factor γ_k characterizes the symmetry of the interaction and of the order parameter. $\gamma_k = 1$, (2.104), which is realized in low-temperature superconductors, corresponds to a gap which opens isotropically on the whole Fermi surface as in the original Bardeen-Cooper-Schrieffer work [60]. This is called a s-wave gap. Our main goal here is to compare the behavior of the isotropic s-wave with the d-wave case, which is relevant for high-Tc superconductors, and it differs from the s-wave for the existence of nodal lines, where the gap amplitude vanishes. In the specific case of the d-wave the nodal lines are the diagonals of the two-dimensional Brillouin zone and the simplest examples are $\gamma_k = \cos k_x - \cos k_y$ on a square lattice and $\gamma_k = k_x^2 - k_y^2$ in a continuum two-dimensional space.

The d-wave symmetry of the order parameter is a very solid experimental evidence, and it arises naturally in approaches where the superconducting pairing is associate with the antiferromagnetic effective interactions arising from superexchange as discussed in Chap. 2 of this manuscript.

In particular, we can easily obtain a d-wave pairing instability in the $t-J$ model [61–63],

$$H = -t\sum_{ij\sigma} P_i c^\dagger_{i\sigma} c_{j\sigma} P_j + J\sum_{ij}(S_i \cdot S_j - \frac{1}{4}) - \mu\sum_i n_i, \qquad (2.107)$$

which is the effective model derived in the $U \gg t$ limit starting from the Hubbard model. The model describes electrons hopping on a square lattice with a hard-core constraint which forbids double occupancy enforced by the projector P_i and interacting via a Heisenberg-like antiferromagnetic double exchange. If we Fourier transform the operators in the Heisenberg interaction we obviously obtain a two-particle term, which can be decoupled in different channels, among which the d-wave channel. It has been repeatedly shown that, in the presence of hopping terms with reasonable dispersions, the d-wave instability is favoured in a wide regions of doping including the relevant regions of the cuprates [61–63]. If we neglect all the other interactions and the hard-core constraint, we obtain explicitly a BCS-like Hamiltonian with $\gamma_k = \cos k_x - \cos k_y$.

In the following discussion we make a further, harmless, approximation by considering a continuum model. This choice allows us to make a direct comparison with an extensive literature on the s-wave BCS model. On the other hand, this somehow defines the simplest and most idealized model for a d-wave superconductor and it is perfectly suited to achieve our goal to disentangle the effects arising from the existence of nodal lines from all the other system-specific properties.

Therefore we consider (2.105) in two dimensions with a free dispersion and a d-wave structure factor

$$\begin{aligned} \varepsilon_k &= t|k|^2 - \mu \\ \gamma_k &= \frac{k_x^2 - k_y^2}{|k|^2} = \cos 2\phi_k, \end{aligned} \qquad (2.108)$$

where μ is the chemical potential, ϕ_k is the polar angle in the plane and $|k|$ is the modulus of the momentum. Since the dispersion ε_k depends only on $|k|$ it can be

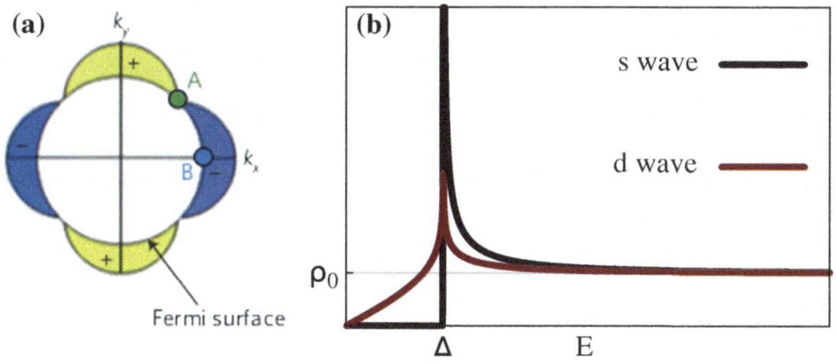

Fig. 2.12 Left: Two dimensional k-space with the circular Fermi surface and the schematic of the d-wave gap, which vanishes at the node (**a**), is maximum at the antinode (**b**) and changes sign by rotation of $\frac{\pi}{2}$ (Fig. from [64]). Right: DOS of quasiparticle excitations. For a fully gapped system, as the s-wave superconductor, the DOS is zero below the gap (black curve). On the other hand, the d-wave superconductor has non-zero DOS for arbitrary low energies (red) (Plot from [59])

useful to parameterize the momentum in terms of the modulus (or equivalently the dispersion) and the angle.

The structure factor clearly vanishes for $k_x = \pm k_y$, or equivalently $\phi_k = (2n + 1)\frac{\pi}{4}$, just like the lattice d-wave factor.

We introduce a momentum cut-off Λ which does not influence the results for weak coupling. It can be practically convenient to impose particle-hole symmetry, which is simply realized by choosing a value of the chemical potential for which the density of states is an even function. This implies $\mu = \frac{t\Lambda^2}{2} = W$.

We do not report here the calculations of the equilibrium mean-field of the BCS model, which can be found in a variety of articles and books. As well known, the main result is that the original quasiparticle dispersion is replaced by a gapped dispersion $E_k = \sqrt{\varepsilon_k^2 + |d_k|^2}$, where $d_k = d\gamma_k$. This implies that, in contrast with the s-wave case, where d_k is a constant in momentum space, a vanishing excitation energy is found when $\varepsilon_k = 0$ and $d_k = 0$, i.e., when the Fermi surface crosses the nodal lines $k_x = \pm k_y$. In other words the superconductor has gapless excitations in the points where the Fermi surface crosses the nodal lines.

The existence of zero- (and low-) energy excitations is reflected in the form of the density of states shown in Fig. 2.12, where the sharp gap of s-wave superconductors is replaced by a linear dependence $D(\varepsilon) \propto \varepsilon$ for small energies, even if sharp features still survive at the gap edges $\pm d$. The existence of nodal lines and low-energy excitations clearly influences the response of the system to external stimuli. In the next chapter we will show that their effect on the non-equilibrium dynamics can be important both from a quantitative and a qualitative point of view.

However, before tackling the direct study of the dynamics, we find it important to review some general properties of the Hamiltonians (2.104) and (2.105) regardless of the isotropic or anisotropic interaction.

2.5.1 Anderson Pseudospin Representation

After the mean-field decoupling, the BCS Hamiltonian is the sum of terms associated with a given momentum. More precisely, fermions with a given momentum and spin "talk" only with fermions with opposite spin and momentum. As a result, the Hilbert space is the direct product of small Hilbert spaces associated with a momentum k which span only four states in the Fock space

$$|0\rangle, |k\uparrow\rangle, |-k\downarrow\rangle, |k\uparrow, -k\downarrow\rangle \tag{2.109}$$

The different subspaces are coupled only via the static mean-field amplitude and there is no pair-breaking term. This implies that the time evolution of each subspace is decoupled from the others and the mean-field can only modify the parameters controlling each individual dynamics, but they can not connect different subspaces.

We can now introduce the so-called Anderson pseudospin operators [65]

$$s_{kx} = \frac{c^\dagger_{k\uparrow}c^\dagger_{-k\downarrow} + c_{-k\downarrow}c_{k\uparrow}}{2}, \tag{2.110a}$$

$$s_{ky} = \frac{c^\dagger_{k\uparrow}c^\dagger_{-k\downarrow} - c_{-k\downarrow}c_{k\uparrow}}{2i}, \tag{2.110b}$$

$$s_{kz} = \frac{c^\dagger_{k\uparrow}c_{k\uparrow} - c_{-k\downarrow}c^\dagger_{-k\downarrow}}{2}, \tag{2.110c}$$

$$s^\pm_k = s_{kx} \pm i s_{ky}. \tag{2.110d}$$

It is easy to verify that the above defined operators satisfy the commutation rules of the angular momentum

$$[s_{ka}, s_{k'b}] = i\delta_{kk'}\varepsilon_{abs}s_{kc} \qquad a, b, c = x, y, z. \tag{2.111}$$

hence they can be treated as spin degrees of freedom.

We can rewrite the mean-field BCS model (2.104) and the definition of the order parameter in terms of the new operators, obtaining

$$H_{\rm MF} = 2\sum_k \varepsilon_k s_{kz} - \sum_{kp}(\mathrm{d}_k s^+_k + \mathrm{d}^*_k s^-_k), \tag{2.112a}$$

$$\mathrm{d}_k = \gamma_k \sum_p \gamma^*_p \langle s^-_p\rangle \tag{2.112b}$$

A very nice feature of the pseudospin representation is that our mean-field Hamiltonian (2.112) commutes with the total spin operators $s^2_k = \langle s^2_{kx}\rangle + \langle s^2_{ky}\rangle + \langle s^2_{kz}\rangle$ (i.e., the sum of the squares of the three components for each momentum), which is therefore a conserved quantity within mean-field. Therefore we can picture the dynamics

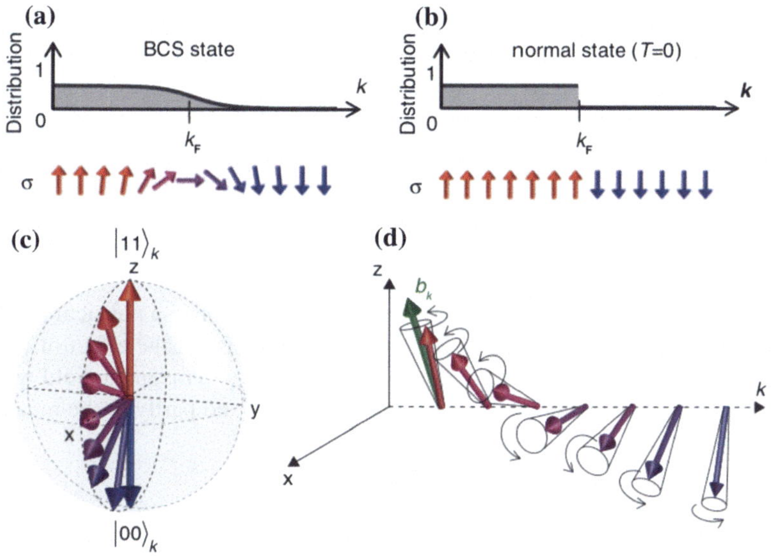

Fig. 2.13 Mapping between the Fermi distribution in momentum space and the Anderson pseudospins in the BCS state (panel A) and in a Fermi liquid (panel B). Panel C shows the Anderson pseudospin on the Bloch sphere while in Panel D the pseudospin precession is pictorially represented (from [66])

in terms of the so-called Bloch sphere, which is the sphere traced by the tip of the arrow representing a spin.

In the pseudospin representation, a completely occupied state in momentum space corresponds to an up pseudospin, while an empty state turns into a down spin. Therefore a simple non-interacting metal has up spins for all momenta below the Fermi momentum and down spin for momenta above it, separated by a sharp edge at the fermionic Fermi surface. In the superconductor instead the opening of the gap translate in smooth rotation of the spin, which now smoothy connects the up region deep inside the occupied region and the down domain deep in the empty region, as depicted in Fig. 2.13).

We briefly mention the connection between the pseudo-spin representation and the Nambu-Gorkov spinor representation. The Nambu spinors are defined as

$$\psi_k = \begin{pmatrix} c_{k\uparrow} \\ c^{\dagger}_{-k\downarrow} \end{pmatrix} \tag{2.113}$$

and we can rewrite the spin operators as

$$s_{ka} = \psi^{\dagger}_k \frac{\boldsymbol{\sigma}}{2} \psi_k. \tag{2.114}$$

In this way the mean-field Hamiltonian (2.105) becomes simply

$$\mathscr{H}_{\mathrm{MF}} = \sum_k \psi_k^\dagger \begin{pmatrix} \varepsilon_k & -\mathrm{d}_k \\ -\mathrm{d}_k^* & -\varepsilon_k \end{pmatrix} \psi_k \tag{2.115}$$

The pseudospin representation is particularly useful because it provides a mapping onto the simple problem of a series of spins (labelled by the momentum) subject to a magnetic field whose z-component is given by the electronic dispersion ε_k and the x- and y-components are given by the mean-field gap.

Moreover, the pseudospin language can be helpful to obtain a physical picture of the superconducting solution, of its dynamics and to consider the effect of external perturbations. In this regard, the pseudospin are subject to a magnetic field which is partly due to the kinetic energy term proportional to ε_k and partly to the mean-field $\dot{s}_k = b_k \times s_k$.

2.5.2 Equations of Motions for the Pairing Amplitudes

In order to study the non-equilibrium dynamics of our model, we can simply introduce the equations of motion for the scalar mean-field amplitudes, which are known as Bloch equations. It is easy to realize that the same equations describe formally the time evolution regardless of the symmetry of the order parameter which is of course included in the structure factor γ_k.

The starting point is the mean-field Hamiltonian

$$H_{\mathrm{MF}} = \sum_{k\sigma} \varepsilon_k c_{k\sigma}^\dagger c_{k\sigma}, - \sum_k (\gamma_k \mathrm{d} c_{k\uparrow}^\dagger c_{-k\downarrow}^\dagger + \gamma_k^* \mathrm{d}^* c_{-k\downarrow} c_{k\uparrow}) \tag{2.116}$$

where d_k is self-consistently determined as

$$\mathrm{d}_k = \lambda \gamma_k \sum_p \gamma_p^* \langle c_{-p\downarrow} c_{p\uparrow} \rangle, \tag{2.117}$$

We can write the equilibrium mean-field equations as

$$G_k = \langle c_{k\sigma}^\dagger c_{k\sigma} \rangle = \frac{1}{2}\left(1 - \frac{\varepsilon_k}{E_k}\right), \tag{2.118}$$

$$F_k = \langle c_{-k\downarrow} c_{k\uparrow} \rangle = \frac{\mathrm{d}_k}{2E_k}, \tag{2.119}$$

where $E_k = \sqrt{\varepsilon_k^2 + \mathrm{d}_k^2}$. G_k is a real number because it is the expectation value of a Hermitian operator while F_k is *a priori* complex but we can choose it to be real without loss of generality.

The equations of motions for G_k and F_k are straightforwardly obtained using the textbook Heisenberg equations of motions for the creation an annihilation operators. After some algebra one obtains

$$i\dot{G}_k = \mathrm{d}_k^* F_k - \mathrm{d}_k F_k^* \tag{2.120}$$

$$i\dot{F}_k = 2\varepsilon_k F_k + \mathrm{d}_k(2G_k - 1) \tag{2.121}$$

where again $\mathrm{d}_k = \gamma_k \sum_p \gamma_p^* F_p$, which implies that also d_k acquires a time dependence through that of F_p.

In the case of s-wave superconductors one can avoid a numerical solution of the Bloch equations because the BCS model is integrable [67]. Without entering the mathematical details, for which we refer to previous literature, the integrability is a consequence of the fact that in the s-wave case the energy dispersion and the interaction depends only on the modulus of the momentum, which makes the problem effectively one-dimensional regardless the original spatial dimensionality of the problem. This leads to a reduction of degrees of freedom and it also allows to use the arsenal of powerful technique designed for one-dimensional problems, like the Bethe ansatz.

This is clearly not possible for any other symmetry of the order parameter since the energy spectrum does not simply depend on the modulus of the momentum, but it has an angular dependence. However, one can design specific combinations of p and d symmetries for which the order parameter does not vanish except for a finite number of points in momentum space. A very popular example is the topological $p + ip$-wave order parameter, with $\gamma_k = k_x + ik_y = |k|e^{i\phi_k}$. Here the excitation energy can vanish at most in one point [68] and the Hamiltonian can be turned into a one-dimensional model by means of the transformation $c_{k\sigma} \to e^{i\phi_k} c_{k\sigma}$. The same idea can be used for the time-reversal breaking order parameter $d + id$-wave for which $\gamma_k = k_x^2 - k_y^2 + 2ik_x k_y = |k|^2 e^{2i\phi_k}$ which also can vanish at most in one point in momentum space [69].

For the standard d-wave case however there is no way to obtain an effective one-dimensional system and we have to resort to a numerical solution of the equations of motion. In the next section we present these results as well as some approximate analytical calculations that we used to interpret and understand our results.

2.5.3 Dynamics After a Quantum Quench

In this chapter we present the results on the dynamics of the gap in a BCS d-wave superconductor after a quantum quench of the interaction [58]. We consider the mean-field Hamiltonian:

$$\mathscr{H}_{\mathrm{MF}} = \sum_{k\sigma} \varepsilon_k c^\dagger_{k\sigma} c_{k\sigma} - \sum_k (\mathrm{d}_k(t) c^\dagger_{k\uparrow} c^\dagger_{-k\downarrow} + \mathrm{d}^*_k c_{-k\downarrow} c_{k\uparrow}), \tag{2.122a}$$

$$\mathrm{d}_k = \lambda \gamma_k \sum_p \gamma^*_p \langle c_{-p\downarrow} c_{p\uparrow} \rangle, \tag{2.122b}$$

and the following configuration for the quantities $G_k(t) = \langle c^\dagger_{k\sigma}(t) c_{k\sigma}(t) \rangle$ and $F_k(t) = \langle c_{-k\downarrow}(t) c_{k\uparrow}(t) \rangle$ that define a superconducting state within mean-field theory:

$$G_k(0) = \frac{1}{2}\left(1 - \frac{\varepsilon_k}{E^{\mathrm{i}}_k}\right), \tag{2.123a}$$

$$F_k(0) = \frac{\mathrm{d}_{\mathrm{i}} \gamma_k}{2 E^{\mathrm{i}}_k}, \tag{2.123b}$$

$$\mathrm{d}_{\mathrm{i}} = \lambda^{\mathrm{i}} \sum_p \gamma^*_p F_k(0), \tag{2.123c}$$

$$E^{\mathrm{i}}_k = \sqrt{\varepsilon^2_K + \mathrm{d}^2_{\mathrm{i}} \gamma^2_k}. \tag{2.123d}$$

Then, we solve the Bloch equations of motion which follow from the Hamiltonian (2.122) with $\lambda^{\mathrm{f}} \neq \lambda^{\mathrm{i}}$:

$$i \dot{G}_k(t) = \gamma_k (\mathrm{d}(t))^* F_k(t) - \gamma_k \mathrm{d}(t) (F_k(t))^*, \tag{2.124a}$$

$$i \dot{F}_k(t) = 2\varepsilon_k F_k(t) + \gamma_k \mathrm{d}(t)(2G_k(t) - 1), \tag{2.124b}$$

$$\mathrm{d}(t) = \lambda^{\mathrm{f}} \sum_p \gamma^*_p F_p(t). \tag{2.124c}$$

The system of (2.124) is a first-order nonlinear system of differential equations for the $2N_s$ quantities $\{G_k(t), F_k(t)\}$ with N_s the number of points in our mesh of the reciprocal space. These equations have to be solved at the same time because they are all coupled to each other via the mean-field $\mathrm{d}(t)$. As discussed previously, for anisotropic models such as ours there is no analytical solution of the problem. That is why we resort to numerical integration of the (2.124) using a Runge-Kutta algorithm at 4th order.

In the following we present the results of this calculation for the cases $\gamma_k = 1$ (s-wave) and $\gamma_k = \cos 2\phi_k$ (d-wave).

2.5.4 The Phase Diagram of s-wave Superductors After a Quantum Quench

The dynamics of an s-wave BCS superconductor after a quantum quench has been the object of a number of investigations motivated by cold-atom experiments [67, 70–72]. As we anticipated above, these works are based on two important properties:

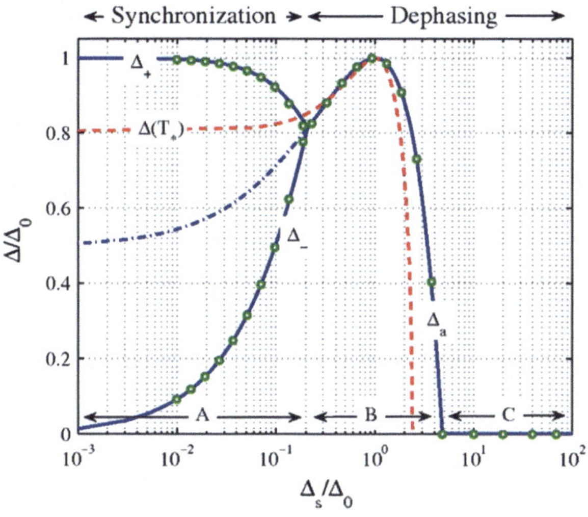

Fig. 2.14 "Phase" diagram for a quantum quench of the interaction of a BCS s-wave superconductor. The horizontal axis is the ratio $\frac{d^i}{d^f}$ (in the figure denoted as $\frac{d_s}{d_0}$). In the regime A (synchronization) the time-dependent gap oscillates indefinitely between the curves d_+ and d_-, while the dotted curve is simply the average. In the two damped regimes it reaches a non-zero (B) or a vanishing (C) stationary value. The red curve represents the thermal value corresponding to the energy pumped into the system via the quench (from [71])

(i) the BCS mean-field Hamiltonian is integrable, i.e., it has as many integrals of motions as degrees of freedom, and (ii) the model is effectively one dimensional because the only dependence on momentum is through its modulus (or equivalently the energy dispersion).

Without entering the details of the theoretical analysis of the integrable system, which lie beyond the scope of this notes, we summarize the "phase diagram" describing the different dynamical regimes of the model which is reported in Fig. 2.14. We discuss the different regimes based on the asymptotic behavior for long times of the time-dependent order parameter.

The dynamics is indeed completely characterized by a single quantity, the ratio $\frac{d^i}{d^f}$ between the the equilibrium gap amplitudes corresponding to the initial value of the coupling (before the interaction quench) and the "final" coupling after the quench. Three regimes have been identified

(A) A region of persistent oscillations, characterized by the synchronization of the different pairing amplitudes for different momenta. In this region the order parameter fluctuates between two different values Δ_+ and Δ_-. In Fig. 2.14 also the average between Δ_+ and Δ_- is reported and compared with $\Delta(T^*)$, the equilibrium gap computed at the temperature T^* corresponding to the energy injected through the quench. This further confirms how this regime is far from thermal.

(B) A regime of damped oscillations. In this region the different pairing amplitudes oscillated, but they acquire a dephasing and the order parameter reaches a finite

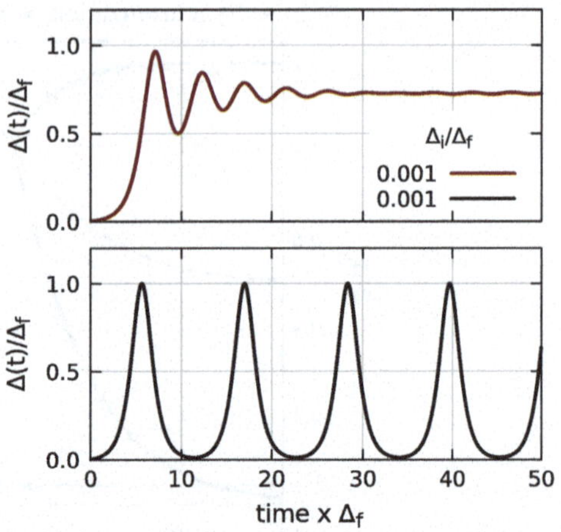

Fig. 2.15 Gap dynamics following a quench from weak to strong interaction $\frac{d_i}{d_f} = 0.001$. In the d-wave superconductor (red) the gap is subject to a fast dephasing after a few oscillations. On the other hand, the s-wave superconductor (black) is in the regime of persistent oscillations (From [59])

asymptotic value. The decay is power-law $\frac{1}{\sqrt{t}}$ in agreement with the pioneering prediction of [73].

(C) An overdamped region, where the dephasing is so strong that the order parameter exponentially decays and rapidly reaches a zero value.

It is important to stress that, even if the gap reaches a stationary value, the system can not relax because of integrability. This means that, also in regimes B and C, the system is not in a thermal state, but it is trapped in a nonthermal state, as confirmed by the disagreement between the asymptotic gaps and $\Delta(T^*)$, which is exactly the gap of the thermal state associated to the quench.

2.5.5 Comparison Between s-wave and d-wave Superconductors

We now compare the nodeless s-wave case with a nodal d-wave superconductor. The latter model is not integrable, therefore we have to resort to a numerical solution of the Bloch equations. We will use the same parametrization of the quench of [71] describing the amplitude of quench in terms of the ratio $\frac{d_i}{d_f}$ which are the equilibrium gap values corresponding to the interaction before the quench λ_i and after the quench λ_f.

In Fig. 2.15 we report the dynamics of the gap after a large quench from weak to strong interaction $\frac{d_i}{d_f} = 0.001$, which lies in the (A) region of the s-wave phase diagram for the s-wave case, as clear from the lower panel, which shows persistent oscillations. We plot the time-dependent order parameter $\Delta(t)$ divided by the "final" value Δ_f which would be the order parameter in an equilibrium system with the after-quench value of the coupling. The contrast with the d-wave case is evident as the d-wave superconductor only shows a mild oscillation which modulates a fast

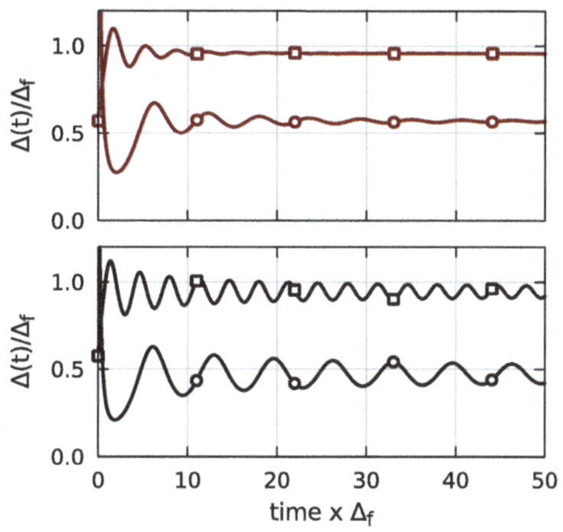

Fig. 2.16 Comparison between d-wave (red) and s-wave (black) dynamics, for quenches of medium strength: $\frac{d_i}{d_f} = 0.5$ (squares) and $\frac{d_i}{d_f} = 3.5$ (circles) (From [59])

relaxation dynamics and the time evolution rapidly converges to a finite asymptotic value, smaller than the equilibrium result. Since the Bloch equations do not allow for dissipation, the vanishing of the gap is clearly associated to the dephasing between the different component at different momenta, while in the s-wave case the different amplitudes retain phase coherence during the time evolution. This means that the individual amplitudes Δ_k do not converge to asymptotic values, but only the k-averaged quantity does.

The disappearance of the regime of persistent oscillations is inevitably associated with the existence of nodal quasiparticles which destroy the momentum-space isotropy of the gap. From a mathematical point of view, the anisotropic gap breaks the integrability, which means that the system can no longer be described merely by a few collective variables, which inhibits the analytical solution that can be obtained for the s-wave case. From a more intuitive point of view, one can loosely associate the breakdown of the coherent oscillations to the existence of low-energy excitations in the nodal regions, which are expected to damp the free oscillations of the gapped particles.

In Fig. 2.16 we report the same comparison between s-wave and d-wave after two different quenches with $\frac{d_i}{d_f} = 0.5$, which falls in the B region of the s-wave superconductor and $\frac{d_i}{d_f} = 3.5$ which is closer to the boundary of the overdamped region C. In this regime both the d- and the s-wave superconductors enter in a regime of damped oscillations and they reach a finite asymptotic value. While the limiting values are not very different in the two cases, the d-wave gap (red) reaches stationarity in shorter times with respect to the s-wave gap (black). This means that, even in the region where the d-wave symmetry does not lead to a qualitative difference with respect to the s-wave case, the existence of nodal quasiparticle excitations favours the dephasing between the pairing amplitudes at different momenta, leading to a faster convergence to the asymptotic value.

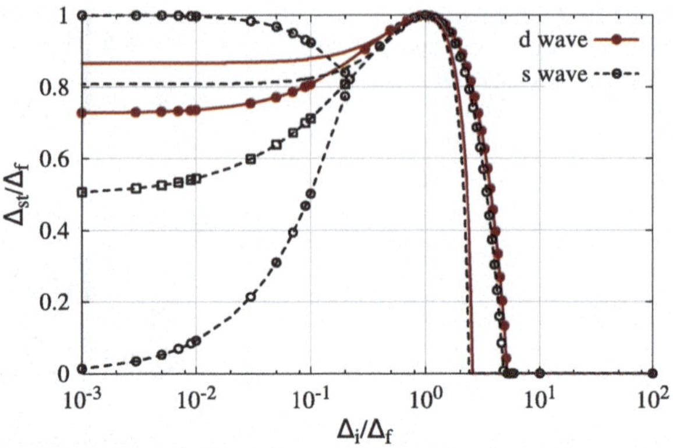

Fig. 2.17 Asymptotic value of the gap $\Delta_{st} = lim_{ft\to\infty}\Delta(t)$ renormalized by Δ_f as a function of the ratio $\frac{\mathrm{d_i}}{\mathrm{d_f}}$ for d-wave (red) and s-wave (black) superconductors after a quantum quench. For the s-wave case and small quenches, the solution oscillates and we plot both the values bounding the oscillations and their average (From [59])

We can summarize our results by superimposing the results for the d-wave symmetry on the phase diagram for the s-wave systems of [71]. It is important to mention that also in the d-wave case, the results only depend on the ratio $\frac{\mathrm{d_i}}{\mathrm{d_f}}$ and not on the actual values of the gaps. Remarkably, the results in the d-wave and s-wave case are quantitatively very close for all quenches where the interaction is reduced ($\frac{\mathrm{d_i}}{\mathrm{d_f}} > 1$) and only small variations are found as long as the s-wave solution displays an actual convergence to an asymptotic value. In the regione where the s-wave superconductor oscillates without damping, the d-wave solution reaches instead an asymptotic value which however does not seem to be correlated with any property of the s-wave solution. We remind however that, even when the asymptotic values of the two symmetries are similar, the characteristic decay time is significantly smaller in the d-wave case (Fig. 2.17).

In order to obtain some analytical insight on the quench dynamics and in particular on the dephasing mechanism, we can consider the limiting case in which the coupling constant vanishes after the quench. This obviously corresponds to $\Delta_f = 0$ and it corresponds therefore to the overdamped region of the phase diagram, where the dephasing is particularly effective.

The advantage of this approximation is simply that during the time evolution at $t > 0$ only the kinetic terms is left and the Hamiltonian is non-interacting so that the field operators evolve in time simply as $c_{k\sigma}(t) = e^{-i\varepsilon_k t} c_{k\sigma}(0)$. Then we can trivially plug the trivial time evolution into the definition of the pairing amplitude $\phi(t) = \sum_k \langle c_{-k\downarrow}(t) c_{k\uparrow}(t)\rangle$. Obviously the time evolution does not depend on the symmetry of the order parameter, which only influences the initial state. We focus therefore on the simplest s-wave case, where we obtain

$$\phi(t) = \sum_k \langle c_{-k\downarrow}(t) c_{k\uparrow}(t)\rangle = \mathrm{d}\rho_0 \int_0^{\frac{W}{\mathrm{d}}} \mathrm{d}x \frac{e^{-2ix\mathrm{d}t}}{\sqrt{x^2+1}} \propto \frac{e^{-2t\mathrm{d}}}{\sqrt{t\mathrm{d}}}, \tag{2.125}$$

where the last approximate relation holds at long times.

2.5.6 Dynamics After a Small Perturbation

We conclude the analysis with another analytical calculation in the specific limit of small quenches. This means that we start from some arbitrary λ_i and we reach a final $\lambda_f = \lambda_i + \delta\lambda$, where $\delta\lambda$ is assumed to be small (but it can be either positive or negative). This means we can treat the quench as small perturbation of the same form of the interaction Hamiltonian proportional to $\delta\lambda$ and use linear response theory to compute the variation of the gap $\delta \mathrm{d}(t) = \mathrm{d}(t) - \mathrm{d_i}$.

The perturbation to the Hamiltonian reads

$$\delta H = \delta\lambda \sum_{kp} \gamma_k \gamma_p^* c_{k\uparrow}^\dagger c_{-k\downarrow}^\dagger c_{-p\downarrow} c_{p\uparrow}. \tag{2.126}$$

using the standard Kubo formula we can calculate the response of the superconducting gap resulting from the external perturbation which is given by

$$\delta d(t) = -i \sum_p \gamma_p^* \int_0^t \mathrm{d}t' \, \langle [c_{-p\downarrow}(t) c_{p\uparrow}(t), \delta\mathscr{H}(t')] \rangle. \tag{2.127}$$

The commutator in (2.127) is easily evaluated if we take a mean-field decoupling which leads to particle-particle and particle-hole terms

$$\delta H = \delta\lambda \sum_p \left[\frac{\mathrm{d}_p}{\lambda^{\mathrm{i}}} (c_{p\uparrow}^\dagger c_{-p\downarrow}^\dagger + c_{-p\downarrow} c_{p\uparrow}) + \gamma_p^2 n_p (c_{p\uparrow}^\dagger c_{p\uparrow} + c_{-p\downarrow}^\dagger c_{-p\downarrow}) \right], \tag{2.128}$$

where n_p and d_p coincide with the unperturbed values of the averages of the density and pairing operators.

After some straightforward, but somewhat involved, manipulation one finds.

$$\delta \mathrm{d}(t) = -\delta\lambda \sum_k \frac{\gamma_k^2 \varepsilon_k^2}{(E_k^{\mathrm{i}})^3} (1 - \cos 2E_k t) . \tag{2.129}$$

Equation (2.129) displays an oscillating behavior, but it has a prefactor where the gapped dispersion appears in the denominator. It is easy to realize that the sum over momenta is dominated by the regions in momentum space where E_k^{i} reaches its minima and by possible singularities in the density of states. For an s-wave superconductor it is well known that the density of states has a square-root divergence at $\pm\Delta_i$. If we take into account only the singular pat of the density of state in (2.129) we obtains a power-law damping superimposed to oscillations with frequency $2\Delta_i$, namely

$$\delta\Delta_s(t)/\Delta_i = \delta\Delta_{s\infty}/\Delta_i + \frac{\delta\lambda\rho_0}{2}\int_{\Delta_i}^{\Lambda} dE \frac{\sqrt{E^2-\Delta_i^2}}{E^2}\cos 2Et \tag{2.130}$$

$$\simeq \delta\Delta_{s\infty}/\Delta_i + A_s\cos(2\Delta_i t+\phi)/t^{\alpha} \tag{2.131}$$

where $\Delta_{s\infty}$ is the asymptotic stationary value of the gap, Λ is a high-energy cut-off, A_s is a constant and ϕ is an initial phase.

In the d-wave case one obtains instead

$$\delta\Delta_d(t)/\Delta_i = \delta\Delta_{d\infty}/\Delta_i + \frac{\pi\delta\lambda}{8v_F v_\Delta \Delta_i^2}\int_0^{\Lambda} dE E^2\cos 2Et \simeq \delta\Delta_{d\infty}/\Delta_i + A_d/t \tag{2.132}$$

where v_F and v_Δ are respectively the Fermi velocity and the nodal velocity that we can extract expanding the gapped dispersion.

The two analytical formulas clearly show how the existence of nodal quasiparticles completely destroys the oscillating contribution to the order parameter and how the relaxation becomes faster in the d-wave case. As a matter of fact the very existence of an asymptotic behavior comes from the destructive interference between the contributions from different momenta (dephasing) which already occurs in the s-wave case, but it is much more effective in the d-wave case.

Acknowledgements The preparation of the lectures and of the lecture notes has greatly benefitted from the reading of the Ph.D. theses by M. Sandri, G. Mazza and F. Peronaci and it is based on the precious collaborations with A. Amaricci, M. Fabrizio, C. Giannetti and M. Schiró on these topics. We also acknowledge useful discussions with A. Avella, V. Brosco, A. Cavalleri, F. Cilento, R. Citro, L. de' Medici, M. Eckstein, D. Fausti, G. Giovannetti, L. Fanfarillo, G. Kotliar, G. Mazza, D. Mihajlovic, F. Novelli, F. Parmigiani, S. Peli, F. Peronaci, G. Sangiovanni, M. Sandri, A. Toschi, A. Valli, P. Werner.

References

1. C. Giannetti, M. Capone, D. Fausti, M. Fabrizio, F. Parmigiani, D. Mihailovic, Ultrafast optical spectroscopy of strongly correlated materials and high-temperature superconductors: a non-equilibrium approach. Adv. Phys. **65**(2), 58–238 (2016)
2. S. Dal Conte, C. Giannetti, G. Coslovich, F. Cilento, D. Bossini, T. Abebaw, F. Banfi, G. Ferrini, H. Eisaki, M. Greven, A. Damascelli, D. van der Marel, F. Parmigiani, Disentangling the electronic and phononic glue in a high-Tc superconductor. Science **335**(6076), 1600–1603 (2012)
3. M. Sentef, A.F. Kemper, B. Moritz, J.K. Freericks, Z.-X. Shen, T.P. Devereaux, Examining electron-boson coupling using time-resolved spectroscopy. Phys. Rev. X **3**, 041033 (2013)
4. S. Dal Conte, L. Vidmar, D. Golez, M. Mierzejewski, G. Soavi, S. Peli, F. Banfi, G. Ferrini, R. Comin, B.M. Ludbrook, L. Chauviere, N.D. Zhigadlo, H. Eisaki, M. Greven, S. Lupi, A. Damascelli, D. Brida, M. Capone, J. Bonca, G. Cerullo, C. Giannetti, Snapshots of the retarded interaction of charge carriers with ultrafast fluctuations in cuprates. Nat. Phys. **11**, 421–426 (2015)

5. B. Mansart, J. Lorenzana, A. Mann, A. Odeh, M. Scarongella, M. Chergui, F. Carbone, Coupling of a high-energy excitation to superconducting quasiparticles in a cuprate from coherent charge fluctuation spectroscopy. Proc. Nat. Acad. Sci. **110**(12), 4539–4544 (2013)
6. F. Novelli, G. De Filippis, V. Cataudella, M. Esposito, I. Vergara, F. Cilento, E. Sindici, A. Amaricci, C. Giannetti, D. Prabhakaran, S. Wall, A. Perucchi, S. Dal Conte, G. Cerullo, M. Capone, A. Mishchenko, M. Grüninger, N. Nagaosa, F. Parmigiani, D. Fausti, Witnessing the formation and relaxation of dressed quasi-particles in a strongly correlated electron system. Nat. Commun. **5**, 5112 (2014)
7. F. Novelli, G. Giovannetti, A. Avella, F. Cilento, L. Patthey, M. Radovic, M. Capone, F. Parmigiani, D. Fausti, Localized vibrations in superconducting $YBa_2Cu_3O_7$ revealed by ultrafast optical coherent spectroscopy. Phys. Rev. B **95**, 174524 (2017)
8. J.D. Rameau, S. Freutel, L. Rettig, I. Avigo, M. Ligges, Y. Yoshida, H. Eisaki, J. Schneeloch, R.D. Zhong, Z.J. Xu, G.D. Gu, P.D. Johnson, U. Bovensiepen, Photoinduced changes in the cuprate electronic structure revealed by femtosecond time- and angle-resolved photoemission. Phys. Rev. B **89**, 115115 (2014)
9. C.L. Smallwood, W. Zhang, T.L. Miller, G. Affeldt, K. Kurashima, C. Jozwiak, T. Noji, Y. Koike, H. Eisaki, D.-H. Lee, R.A. Kaindl, A. Lanzara, Influence of optically quenched superconductivity on quasiparticle relaxation rates in $Bi_2Sr_2CaCu_2O_{8+\delta}$. Phys. Rev. B **92**, 161102 (2015)
10. C. Giannetti, F. Cilento, S.D. Conte, G. Coslovich, G. Ferrini, H. Molegraaf, M. Raichle, R. Liang, H. Eisaki, M. Greven, A. Damascelli, D. van der Marel, F. Parmigiani, Revealing the high-energy electronic excitations underlying the onset of high-temperature superconductivity in cuprates. Nat. Commun. **2**, 353 (2011)
11. F. Cilento, S. Dal Conte, G. Coslovich, S. Peli, N. Nembrini, S. Mor, F. Banfi, G. Ferrini, H. Eisaki, M.K. Chan, C.J. Dorow, M.J. Veit, M. Greven, D. van der Marel, R. Comin, A. Damascelli, L. Rettig, U. Bovensiepen, M. Capone, C. Giannetti, F. Parmigiani, Photo-enhanced antinodal conductivity in the pseudogap state of high-Tc cuprates. Nat. Commun. **5**, 4353 (2014)
12. S. Peli, S.D. Conte, R. Comin, N. Nembrini, A. Ronchi, P. Abrami, F. Banfi, G. Ferrini, D. Brida, S. Lupi, M. Fabrizio, A. Damascelli, M. Capone, G. Cerullo, C. Giannetti, Mottness at finite doping and charge instabilities in cuprates. Nat. Phys. **13**, 806–811 (2017)
13. D. Polli, M. Rini, S. Wall, R.W. Schoenlein, Y. Tomioka, Y. Tokura, G. Cerullo, A. Cavalleri, Coherent orbital waves in the photo-induced insulator-metal dynamics of a magnetoresistive manganite. Nat. Mater. **6**, 643 (2007)
14. D. Fausti, R.I. Tobey, N. Dean, S. Kaiser, A. Dienst, M.C. Hoffmann, S. Pyon, T. Takayama, H. Takagi, A. Cavalleri, Light-induced superconductivity in a stripe-ordered cuprate. Science **331**(6014), 189–191 (2011)
15. R. Mankowsky, A. Subedi, M. Forst, S.O. Mariager, M. Chollet, H.T. Lemke, J.S. Robinson, J.M. Glownia, M.P. Minitti, A. Frano, M. Fechner, N.A. Spaldin, T. Loew, B. Keimer, A. Georges, A. Cavalleri, Nonlinear lattice dynamics as a basis for enhanced superconductivity in $YBa_2Cu_3O_{6.5}$. Nature **516**, 71–73 (2014)
16. T. Langen, R. Geiger, J. Schmiedmayer, Ultracold atoms out of equilibrium. Ann. Rev. Condens. Matter Phys. **6**(1), 201–217 (2015)
17. A. Polkovnikov, K. Sengupta, A. Silva, M. Vengalattore, Colloquium. Rev. Mod. Phys. **83**, 863–883 (2011)
18. U. Schollwöck, The density-matrix renormalization group. Rev. Mod. Phys. **77**, 259–315 (2005)
19. A. Georges, G. Kotliar, W. Krauth, M.J. Rozenberg, Dynamical mean-field theory of strongly correlated fermion systems and the limit of infinite dimensions. Rev. Mod. Phys. **68**, 13–125 (1996)
20. H. Aoki, N. Tsuji, M. Eckstein, M. Kollar, T. Oka, P. Werner, Nonequilibrium dynamical mean-field theory and its applications. Rev. Mod. Phys. **86**, 779–837 (2014)
21. G. Kotliar, A.E. Ruckenstein, New functional integral approach to strongly correlated fermi systems: the gutzwiller approximation as a saddle point. Phys. Rev. Lett. **57**, 1362–1365 (1986)

22. F. Lechermann, A. Georges, G. Kotliar, O. Parcollet, Rotationally invariant slave-boson formalism and momentum dependence of the quasiparticle weight. Phys. Rev. B **76**, 155102 (2007)
23. A. Isidori, M. Capone, Rotationally invariant slave bosons for strongly correlated superconductors. Phys. Rev. B **80**, 115120 (2009)
24. S. Florens, A. Georges, Slave-rotor mean-field theories of strongly correlated systems and the Mott transition in finite dimensions. Phys. Rev. B **70**, 035114 (2004)
25. L. de' Medici, M. Capone, *Modeling Many-Body Physics with Slave-Spin Mean-Field: Mott and Hund's Physics in Fe-Superconductors* (Springer International Publishing, Cham, 2017), pp. 115–185
26. N.F. Mott, The basis of the electron theory of metals, with special reference to the transition metals. Proc. Phys. Soc. Sect. A **62**(7), 416 (1949)
27. E.H. Lieb, F.Y. Wu, Absence of Mott transition in an exact solution of the short-range, one-band model in one dimension. Phys. Rev. Lett. **20**, 1445–1448 (1968)
28. A. Toschi, M. Capone, C. Castellani, Energetic balance of the superconducting transition across the BCS-Bose Einstein crossover in the attractive Hubbard model. Phys. Rev. B **72**(23), 235118 (2005)
29. M. Keller, W. Metzner, U. Schollwck, Dynamical mean-field theory for pairing and spin gap in the attractive Hubbard model. Phys. Rev. Lett. **86**, 4612–4615 (2001)
30. M. Capone, C. Castellani, M. Grilli, First-order pairing transition and single-particle spectral function in the attractive Hubbard model. Phys. Rev. Lett. **88**(12), 126403 (2002)
31. A. Toschi, P. Barone, M. Capone, C. Castellani, Pairing and superconductivity from weak to strong coupling in the attractive Hubbard model. New J. Phys. **7**(ii), 7–7 (2005)
32. G. Mazza, From sudden quench to adiabatic dynamics in the attractive hubbard model. arXiv:1708.01096 (2017)
33. A. Amaricci, M. Capone, Dynamical mean-field theory description of the voltage-induced transition in a nonequilibrium superconductor. Phys. Rev. B **93**, 014508 (2016)
34. M.C. Gutzwiller, Effect of correlation on the ferromagnetism of transition metals. Phys. Rev. **134**, A923–A941 (1964)
35. M.C. Gutzwiller, Correlation of electrons in a narrow s band. Phys. Rev. **137**, A1726–A1735 (1965)
36. W.F. Brinkman, T.M. Rice, Application of gutzwiller's variational method to the metal-insulator transition. Phys. Rev. B **2**, 4302–4304 (1970)
37. M. Schiró, M. Fabrizio, Time-dependent mean field theory for quench dynamics in correlated electron systems. Phys. Rev. Lett. **105**(7), 076401 (2010)
38. M. Schiró, M. Fabrizio, Quantum quenches in the Hubbard model: time-dependent mean-field theory and the role of quantum fluctuations. Phys. Rev. B **83**, 165105 (2011)
39. M. Fabrizio, *The Out-of-Equilibrium Time-Dependent Gutzwiller Approximation* (Springer Netherlands, Dordrecht, 2013), pp. 247–273
40. M. Sandri, The Gutzwiller Approach to out-of-equilibrium correlated fermions. Ph.D. thesis, SISSA (October 2014)
41. G. Mazza, Non Equilibrium Phenomena in Strongly Correlated Systems. Ph.D. thesis, SISSA (2014/2015)
42. G. Seibold, J. Lorenzana, Time-dependent gutzwiller approximation for the Hubbard model. Phys. Rev. Lett. **86**, 2605–2608 (2001)
43. J. Bünemann, M. Capone, J. Lorenzana, G. Seibold, Linear-response dynamics from the time-dependent gutzwiller approximation. New J. Phys. **15**(5), 053050 (2013)
44. G. Mazza, A. Amaricci, M. Capone, M. Fabrizio, Electronic transport and dynamics in correlated heterostructures. Phys. Rev. B **91**, 195124 (2015)
45. M. Sandri, M. Fabrizio, Nonequilibrium gap collapse near a first-order Mott transition. Phys. Rev. B **91**, 115102 (2015)
46. A. Georges, Strongly correlated electron materials: dynamical mean-field theory and electronic structure. AIP Conf. Proc. **715**(1), 3–74 (2004)

47. D. Vollhardt, K. Byczuk, M. Kollar, *Dynamical Mean-Field Theory* (Springer, Berlin, 2012), pp. 203–236
48. W. Metzner, D. Vollhardt, Correlated lattice fermions in $d = \infty$ dimensions. Phys. Rev. Lett. **62**, 324–327 (1989)
49. A. Georges, G. Kotliar, Hubbard model in infinite dimensions. Phys. Rev. B **45**, 6479–6483 (1992)
50. M. Eckstein, M. Kollar, P. Werner, Thermalization after an interaction quench in the Hubbard model. Phys. Rev. Lett. **103**, 056403 (2009)
51. M. Moeckel, S. Kehrein, Interaction quench in the Hubbard model. Phys. Rev. Lett. **100**, 175702 (2008)
52. M. Moeckel, S. Kehrein, Crossover from adiabatic to sudden interaction quenches in the hubbard model: prethermalization and non-equilibrium dynamics. New J. Phys. **12**(5), 055016 (2010)
53. M. Kollar, F.A. Wolf, M. Eckstein, Generalized Gibbs ensemble prediction of prethermalization plateaus and their relation to nonthermal steady states in integrable systems. Phys. Rev. B **84**, 054304 (2011)
54. M. Eckstein, M. Kollar, Near-adiabatic parameter changes in correlated systems: influence of the ramp protocol on the excitation energy. New J. Phys. **12**(5), 055012 (2010)
55. M. Sandri, M. Schiró, M. Fabrizio, Linear ramps of interaction in the fermionic Hubbard model. Phys. Rev. B **86**, 075122 (2012)
56. N. Tsuji, M. Eckstein, P. Werner, Nonthermal antiferromagnetic order and nonequilibrium criticality in the Hubbard model. Phys. Rev. Lett. **110**, 136404 (2013)
57. M. Sandri, M. Fabrizio, Nonequilibrium dynamics in the antiferromagnetic Hubbard model. Phys. Rev. B **88**, 165113 (2013)
58. F. Peronaci, M. Schiró, M. Capone, Transient dynamics of d-wave superconductors after a sudden excitation. Phys. Rev. Lett. **115**, 257001 (2015)
59. F. Peronaci, Transient Dynamics of Unconventional Superconductors: d-wave Symmetry and Strong Correlations. Ph.D. thesis, SISSA (October 2016)
60. J. Bardeen, L.N. Cooper, J.R. Schrieffer, Theory of superconductivity. Phys. Rev. **108**, 1175–1204 (1957)
61. G. Kotliar, Resonating valence bonds and d-wave superconductivity. Phys. Rev. B **37**, 3664–3666 (1988)
62. G. Kotliar, J. Liu, Superexchange mechanism and d-wave superconductivity. Phys. Rev. B **38**, 5142–5145 (1988)
63. P.A. Lee, N. Nagaosa, X.-G. Wen, Doping a mott insulator: physics of high-temperature superconductivity. Rev. Mod. Phys. **78**, 17–85 (2006)
64. M. Hashimoto, I.M. Vishik, R.-H. He, T.P. Devereaux, Z.-X. Shen, Energy gaps in high-transition-temperature cuprate superconductors. Nat. Phys. **10**, 483–495 (2014)
65. P.W. Anderson, Random-phase approximation in the theory of superconductivity. Phys. Rev. **112**, 1900–1916 (1958)
66. R. Matsunaga, N. Tsuji, H. Fujita, A. Sugioka, K. Makise, Y. Uzawa, H. Terai, Z. Wang, H. Aoki, R. Shimano, Light-induced collective pseudospin precession resonating with Higgs mode in a superconductor. Science **345**(6201), 1145–1149 (2014)
67. E.A. Yuzbashyan, B.L. Altshuler, V.B. Kuznetsov, V.Z. Enolskii, Nonequilibrium cooper pairing in the nonadiabatic regime. Phys. Rev. B **72**, 220503 (2005)
68. M.S. Foster, V. Gurarie, M. Dzero, E.A. Yuzbashyan, Quench-induced floquet topological p-wave superfluids. Phys. Rev. Lett. **113**, 076403 (2014)
69. I. Marquette, J. Links, Integrability of an extended $d + id$-wave pairing hamiltonian. Nucl. Phys. B **866**(3), 378–390 (2013)
70. R.A. Barankov, L.S. Levitov, B.Z. Spivak, Collective Rabi oscillations and solitons in a time-dependent BCS pairing problem. Phys. Rev. Lett. **93**, 160401 (2004)
71. R.A. Barankov, L.S. Levitov, Synchronization in the BCS pairing dynamics as a critical phenomenon. Phys. Rev. Lett. **96**, 230403 (2006)

72. E.A. Yuzbashyan, O. Tsyplyatyev, B.L. Altshuler, Relaxation and persistent oscillations of the order parameter in fermionic condensates. Phys. Rev. Lett. **96**, 097005 (2006)
73. A.F. Volkov, S.M. Kogan, Collisionless relaxation of the energy gap in superconductors. Sov. J. Exp. Theor. Phys. **38**, 1018 (1974)

Chapter 3
From the Keldysh Formalism to Non-equilibrium Dynamical Mean-Field Theory

Martin Eckstein

Abstract Using femto-second laser-pulses one can investigate correlated electrons in solids far from equilibrium, and possibly even manipulate their behavior in a controlled and ultrafast manner. The theoretical understanding of such non-equilibrium situations meets considerable challenges, as established concepts like quasiparticles are rigorously defined only in the limit of low-energy equilibrium states. In these notes we discuss the foundations of non-equilibrium dynamical mean-field theory, an approach which allows to investigate the dynamics of correlated electron systems on microscopic timescales. We start from the definition of real-time Green's functions, which provide a rigorous framework to interpret electronic structure out of equilibrium. We then introduce the Keldysh formalism and its relation to the description of non-equilibrium states in terms of kinetic equations. Finally we discuss non-equilibrium dynamical mean-field theory and some of its applications. We focus on photo-induced states in Mott insulators, which provides a paradigm example for a non-equilibrium system where well-defined quasiparticles are not established.

3.1 Introduction

A variety of complex condensed matter phases arise from the subtle interplay between electronic charge, spin, and orbital degrees of freedom [1, 2]. Paradigm examples are given by transition-metal compounds, which often show magnetic, orbital, and charge-order in close vicinity, by the finding of high-temperature superconductivity in cuprates and iron-pnictides, or by heavy-fermion materials. These systems often react in a highly sensitive manner to external parameters, which implies technologically relevant phenomena like the colossal magnetoresistance. Since the early days of condensed matter physics it has been a major goal to unravel the origin of such states, and to understand how they can be controlled by external fields. For a long time, this area of research has been in the realm of equilibrium physics, i.e., external

M. Eckstein (✉)
Friedrich-Alexander Universität Erlangen-Nürnberg, Erlangen, Germany
e-mail: martin.eckstein@fau.de

R. Citro and F. Mancini (eds.), *Out-of-Equilibrium Physics of Correlated Electron Systems*, Springer Series in Solid-State Sciences 191,
https://doi.org/10.1007/978-3-319-94956-7_3

control parameters like pressure or magnetic fields vary slow enough so that the system remains in equilibrium on the microscopic scale. This has changed with the experimental possibilities which are opened up by ultrafast laser spectroscopy. Short laser pulses are nowadays available ranging from highly intense terahertz transients of picosecond duration and peak-fields of volts per lattice constant, to femtosecond pulses in the infra-red, visible, or even X-ray domain. These pulses can be used to excite materials and probe their dynamics before the return to equilibrium [3]. From the analysis of the relaxation after a perturbation one can get a new view on the underlying degrees of freedom and their interaction. Examples include the characterization of the superconducting pairing interaction in the time domain [4], or the distinction of charge-density waves of lattice and electronic origin by how they melt after a photo-excitation [5]. However, one can go further than this, and drive a system so far out of equilibrium that its behavior becomes qualitatively different from the states in the equilibrium phase diagram. For example, nonlinear phononics allows to change the lattice constants and the corresponding electronic properties by strongly driving an-harmonically coupled optical phonons [6], signatures of light-induced superconducting response have been observed in cuprates [7, 8] and fullerides [9] during stimulation in the multi terahertz regime, and one can even switch to hidden states, i.e., long-lived states with new types of magnetic and orbital order which are accessible only via ultra-short excitation [10, 11].

From the theoretical perspective, the understanding of such non-thermal states in solids is linked to one of the most fundamental questions of statistical physics, i.e., how fast and whether an interacting system would *thermalize* after a perturbation. A laser can easily deposit an energy corresponding to several thousand Kelvin into the electronic system. In a thermal equilibrium state at such a high temperature, we do not expect well-defined quasiparticles nor long-range electronic order of superconducting, magnetic, or any other type. If photo-excitation of a system should *enhance* an order parameter (as observed in recent experiments on excitonic insulators [12]), the system must not thermalize on relevant timescales. The thermalization of many-body quantum systems has been extensively investigated in the controlled setup of ultra-cold gas experiments, and it would require a separate chapter to discuss these issues here [13]. There are indeed ideal model systems which strictly do not behave ergodic, in particular many-body localized and integrable ones. Examples of many-body localized systems have so far not been identified in condensed matter, and integrable Hamiltonians are not generic. Yet, even when a system is only close to an integrable point, its behavior on early times can reflect the non-ergodic behavior of the integrable system. The resulting behavior is a two-stage relaxation consisting of fast prethermalization followed by a subsequent thermalization [14–16]. Prethermalized states may show long-range order even if the corresponding thermal state does not [17–19]. Moreover, even far from integrable points, fast electronic thermalization is not obvious, as we will see in these notes for the case of electrons close to the Mott transition. This leaves a wide area for the discovery of novel long-lived non-equilibrium phases.

This new area of condensed matter physics puts great challenges to theory, as many fundamental concepts of condensed mater physics break down on the relevant timescales. An important example is the notion of electronic quasiparticles. An intuitive picture of a non-equilibrium state is that of a gas of quasiparticles whose dynamics can be described in terms of a Boltzmann equation. The Boltzmann equation is a classical rate-equation for the time-evolution of a phase-space density, i.e., the occupation probability of momentum and position states. It has been used with great success, e.g., to describe the dynamics of Fermi liquids in near-equilibrium situations, including transport and collective oscillations [20]. However, quasiparticles are only well-defined close to the ground state, and often become fragile in complex materials, in particular close to metal insulator phase transitions or quantum phase transitions with diverging fluctuation length. Moreover, even the familiar concept of a single-particle band-structure is an equilibrium concept, as the bands depend the screened lattice-periodic potential, which is determined by the collective response of the electronic system itself. Similarly, all interactions in the solid, from the screened electron-electron interaction to emergent interactions such as superconducting pairing and spin-exchange, are subject to non-equilibrium excitations.

A viable approach to non-equilibrium many-body systems is the Keldysh formalism, which was developed in the 1960s by Keldsyh, Kadanoff, Baym, Schwinger, and others [21, 22]. The Keldysh formalism is based on Green's functions $G(\boldsymbol{r}, t, \boldsymbol{r}', t')$, which describe the propagation of particles and holes between space-time point $\boldsymbol{r}, t$ and $\boldsymbol{r}', t'$. These propagators contain both the information on the spectral function (electronic structure, quasiparticle energies) and the occupation of states, which mutually depend on each other. Only on the level of mean-field dynamics (such as in a time-dependent BCS theory for superconductors), and within the kinetic equation approach, one can rigorously separate the two concepts in terms of the eigenvalues of a single-particle mean-field Hamiltonian and the corresponding one-particle density matrix. However, present experiments are performed at timescales and excitation densities where one cannot rely on an approximate separation of spectral and occupation functions. Calculations based on two-time Green's functions are noways possible numerically, and have been applied to a wide range of systems, from the dynamics of the screening in semiconductors [23], to applications in nuclear physics, cosmology, and plasma physics [24], as well as in inhomogeneous systems such as small Coulomb clusters or molecules [25]. More recent applications include perturbative simulations of the dynamics in various lattice models, such as spins in the Heisenberg model [26], the study of photo-excited states in electron-phonon systems using time-dependent Migdal-Eliashberg theory [27, 28], or time-dependent GW simulations of excitonic condensates [19]. Finally, the reformulation of dynamical mean-field theory (DMFT) [29] in the Keldysh formalism, which is the main subject of these notes, allowes to study the dynamics of strongly correlated systems.

These notes aim to present the basic notions for a theory of correlated electronic states out of equilibrium. Section 3.2 covers some elementary concepts, i.e., the definition of equilibrium and non-equilibrium states in terms of the Gibbs ensemble and the fluctuation dissipation relation (Sect. 3.2.1), and the definition of spectral functions and quasiparticles in terms of Green's functions (Sect. 3.2.2). In Sect. 3.3 and 3.4 we introduce the Keldysh formalism, and discuss the solution of the Dyson equation in real time, which can be considered as a generalized Boltzmann equation with a memory kernel. Finally, in Sect. 3.5 we introduce the formalism of non-equilibrium dynamical mean-field theory, and apply it in Sect. 3.6 to study photo-induced states in Mott insulators.

3.2 Green's Functions and Many-Body Systems Out of Equilibrium

In this chapter, we introduce the notion of thermal equilibrium, as well as that of single-particle excitations and Green's functions. These are basic concepts of statistical physics and many-body physics, and there is of course extensive literature on this topic [20]. For most readers this section will therefore not provide much new, but it may still be illustrating to recapitulate such known concepts from a perspective which aims at the description of non-equilibrium states from the beginning.

3.2.1 The Thermal Equilibrium State

The Gibbs Ensemble and the Eigenstate Thermalization Hypothesis

In the language of statistical physics, the thermal equilibrium state is described by the Gibbs ensemble with a density matrix

$$\rho = \frac{1}{Z}\sum_m e^{-\beta E_m} |m\rangle\langle m|, \tag{3.1}$$

in which the statistical weight of an energy eigenstate $|m\rangle$ with energy E_m is given by the Boltzmann factor $e^{-\beta E_m}/Z$ with inverse temperature $\beta = 1/k_B T$. We will use the Gibbs ensemble to describe the initial state of a system before it is excited, and we will compare the time-evolving state of a system to the Gibbs ensemble in order to access to what extent it is thermalized.

Now suppose we would like to decide, by some kind of measurement, whether a system at hand is in thermal equilibrium. Does this mean the probability of finding it in an eigenstate $|m\rangle$ should be given by $e^{-\beta E_m}/Z$? The answer is no. In fact, the notion of thermal equilibrium is useful because of its *universality*, which means that a vast number of states are indistinguishable from the density matrix (3.1) by a reasonable physical measurement, i.e., a measurement of a local or few-particle observable. It is also only within this notion of indistinguishability that we can believe that a generic system of interacting particles in the thermodynamic limit would relax to the Gibbs ensemble when it is isolated from the environment, even though its quantum state remains pure. What is meant by a "vast number of states" is formalized by the eigenstate thermalization hypothesis (ETH), which conjectures that in the thermodynamic limit the expectation value $\langle m|\mathcal{O}|m\rangle$ of a local observable $\mathcal{O}$ in *any eigenstate* $|m\rangle$ with an energy close to the average energy of (3.1) approaches the expectation value in the microcanonical (or canonical) ensemble [30–32]. In other words, for systems which satisfy ETH, each state $|m\rangle$ is a suitable representative of the ensemble, and thus also a random superposition of eigenstates. In contrast, for systems which do not satisfy ETH, the Gibbs ensemble is probably not a very useful concept, as a local measurement would lead to a different outcome for each realization of the ensemble. The ETH is the quantum analog of ergodicity in classical mechanics, which states that the whole phase space is sampled by the time-average along a single trajectory.

The Fluctuation-Dissipation Theorem

According to the previous discussion, we can verify whether a system is in thermal equilibrium by comparing local or few-body observables to the prediction from a Gibbs ensemble with a suitable temperature. This need of an external thermal reference state is clearly unsatisfactory, and one should rather find a way to see whether or not a system is in equilibrium by looking at the properties of the system alone. If a system consists of independent particles, like a gas of free fermions, we can trade the un-measurable many-body occupation probabilities in (3.1) for single-particle occupations: In thermal equilibrium, a single-particle energy level with energy ϵ should be occupied with the Fermi or Bose distribution $1/(e^{\beta\epsilon} \pm 1)$. This criterion, however, becomes at most approximate, and often useless, as no system in condensed matter is truly described by independent particles. Instead, in many-body physics, we describe a system by its elementary excitations, defined by some general field $X(\boldsymbol{r})$ (e.g., the displacement field of atoms in a crystal). The *response* of X to an external force defines the *spectrum* of the excitations, while the statistical and quantum fluctuations of X correspond the occupation of the elementary excitations. As we will see now, thermal equilibrium is characterized by a fundamental relation between spectrum and occupation, or response and fluctuations, the *fluctuation dissipation theorem*.

In a general linear response setting we consider a Hamiltonian $H = H_0 + h(t)X$ which is slightly perturbed from H_0 by an external field h that couples through an operator X to the system. Using time-dependent perturbation theory, we can obtain the change of the expectation value of X itself under the influence of the field. To leading order in h (assuming for simplicity that the equilibrium expectation value of X vanishes), the response is given in terms of a retarded response kernel χ (Kubo formula [33]),

$$\langle X(t)\rangle = \int ds\ \chi(t,s)h(s), \tag{3.2}$$

$$\chi(t-t') = -i\theta(t-t')\langle[X(t), X(t')]\rangle. \tag{3.3}$$

In the expression for χ, the time-dependence of X is computed with respect to the undisturbed Hamiltonian H_0, and the brackets denote the initial equilibrium state $\langle\cdots\rangle = \mathrm{tr}[e^{-\beta H_0}\cdots]/Z$. On the other hand, one can consider the autocorrelation functions

$$C^>(t-t') = -i\langle X(t)X(t')\rangle, \tag{3.4}$$
$$C^<(t-t') = -i\langle X(t')X(t)\rangle. \tag{3.5}$$

Since in equilibrium all correlation functions depend on time-difference only, it is convenient to define Fourier transform, $\chi(\omega) = \int dt\, \chi(t)e^{i\omega t}$ and $C(\omega) = \int dt\, C(t)e^{i\omega t}$. On can now represent the $C(\omega)$ and $\chi(\omega)$ using an eigenbasis $|m\rangle$ of H_0 (the so-called Lehmann representation)

$$\begin{aligned} C^<(\omega) &= -\frac{i}{Z}\sum_{n,m} e^{-\beta E_n}|\langle n|X|m\rangle|^2 \int dt\, e^{i\omega t}e^{it(E_m-E_n)} \\ &= -\frac{2\pi i}{Z}\sum_{n,m} e^{-\beta E_n}|\langle n|X|m\rangle|^2\delta(\omega - E_n + E_m). \end{aligned} \tag{3.6}$$

Similarly, we have

$$\begin{aligned} \chi(\omega+i0) &= -\frac{i}{Z}\sum_{n,m} e^{-\beta E_n}|\langle n|X|m\rangle|^2 \int_0^\infty dt\, e^{i(\omega+i0)t}[e^{it(E_n-E_m)} - e^{-it(E_n-E_m)}], \\ \Rightarrow \ \mathrm{Im}\chi(\omega+i0) &= -\frac{\pi}{Z}\sum_{n,m}[e^{-\beta E_n} - e^{-\beta E_m}]|\langle n|X|m\rangle|^2\delta(\omega + E_n - E_m). \end{aligned} \tag{3.7}$$

When we use the δ-function to replace $e^{-\beta E_m}$ by $e^{-\beta E_n}e^{-\beta\omega}$ in (3.6) and (3.7), we obtain the fundamental relation

$$C^<(\omega) = 2ib(\omega)\,\mathrm{Im}\chi(\omega+i0), \tag{3.8}$$

where $b(\omega) = \frac{1}{e^{\beta\omega}-1}$ is the Bose function. This is the fluctuation dissipation theorem [33], which states that the imaginary part of the response function, which describes energy dissipation, is related to the power spectrum of the fluctuations.[1] It holds for any linear response relation, and paradigm examples include the Nyquist noise in a resistor, and the Einstein relation between the diffusion and friction constants. In terms of elementary excitations, it implies that *occupation function* $C^<(\omega)$ of excitations is given by the spectrum $\mathrm{Im}\chi(\omega+i0)$ and the Bose function, in a analogy

[1]Usually, the fluctuation dissipation theorem is stated in terms of the symmetrized autocorrelation function $C = i(C^< + C^>)$, for which it reads $C(\omega) = -2\coth\left(\frac{\beta\omega}{2}\right)\mathrm{Im}\chi(\omega+i0)$..

to the independent particle case.[2] The analogous case of fermionic single-particle excitation is presented in the next section. Hence quantum field theory in equilibrium can be developed solely in terms of the response propagators or spectral functions (or their analytical continuation to imaginary frequency), while quantum field theory in non-equilibrium must provide independent equations for both response and fluctuations. This is the Keldysh formalism, which will be introduced in Sect. 3.3.

Two-Time Response Functions

In a time-evolving state time-translational invariance is lost, so that both response and correlation functions depend on two time arguments separately. It is still often convenient to introduce a partial frequency dependence: A suitable symmetric choice is the *Wigner transform* for a function $F(t, t')$, where we introduce a center of mass time (or average time) $t_{av} = (t + t')/2$ and a relative time $t_{rel} = t - t'$, and perform a Fourier transform with respect to t_{rel}

$$F(t_{av}, \omega) = \int dt_{rel}\, e^{i\omega t_{rel}}\, F(t_{av} + t_{rel}/2, t_{av} - t_{rel}/2). \tag{3.9}$$

In equilibrium, there is no dependence on average time. In non-equilibrium, we can define a response function $\chi(t, \omega)$, a correlation function $C^<(t, \omega)$, and, in analogy to (3.8), a distribution function

$$h(t, \omega) = \frac{C^<(t, \omega)}{2i\, \mathrm{Im}\chi(t, \omega + i0)}. \tag{3.10}$$

Verifying the fluctuation-dissipation theorem by showing that $h(t, \omega)$ equals the Bose function in principle provides a way to determine whether a system is in equilibrium or not. Unfortunately, measuring fluctuations of a quantity is generally much harder than measuring the response. For example, optical pump-probe experiments measure the dielectric function $\epsilon(t, \omega)$, which is related to the polarization response to long-wavelength electric fields. According to the discussion above, a full characterization of the non-equilibrium state would require also to measure the polarization or current noise. Recent experiments in this direction have been performed, which extract current fluctuations from the shot-to-shot variance in the measured intensity of the reflected probe-pulse in a pump-probe experiment [34].

It should be noted that in order to interpret $h(t, \omega)$ as an occupation of states, it should at least be positive for $\omega > 0$, which is not true in general for (3.10). This issue will be discussed in Sect. 3.2.3. A special situation is a *non-equilibrium steady state*, i.e., where all properties are translationally invariant in time, but nevertheless the system does not satisfy the fluctuation-dissipation theorem. This situation may

---

[2]For elementary excitations which are characterized by momentum $\boldsymbol{k}$, one can simply take X in (3.2) to be the corresponding Fourier component $X_{\boldsymbol{k}}$ of the field $X(\boldsymbol{r})$ and add momentum indices throughout all equations.

occur in a non-ergodic system which does not thermalize, but more generic are non-equilibrium steady states in open dissipative systems: If a small system, which is coupled to an infinite bath of states, is subject to a steady perturbation, a non-equilibrium situation can emerge in which the absorption of energy or particles from the perturbation is balanced by a corresponding flow into the environment. When correlation functions do not depend on average time, we can show that the spectrum and the occupation function $h(\omega)$ are indeed positive for $\omega > 0$ (see Sect. 3.2.3). In this case, the deviation of the occupation $h(\omega)$ from a Bose function indicates the deviation from a non-equilibrium state.

3.2.2 Green's Functions and Electronic Structure

In equilibrium, the electronic structure is fully characterized by the spectral function, which defines the electronic bands in an interacting system. In this section, we discuss the generalization of the spectral function to non-equilibrium situations in terms of *two-time Green's functions*, which have a close link to the response and correlation functions of the previous section. It is convenient to start from the following electron and hole propagators, as introduced by Kadanoff and Baym [22],

$$G^{<}_{a,a'}(t,t') = +i\langle c^{\dagger}_{a'}(t')c_a(t)\rangle \tag{3.11}$$

$$G^{>}_{a,a'}(t,t') = -i\langle c_{a'}(t)c^{\dagger}_{a}(t')\rangle. \tag{3.12}$$

Here $c^{\dagger}_{a}$ and c_a denote creation and annihilation operators for an electron in a single-particle orbital $a \in \{1, ..., L\}$ (which may denote spin, momentum, position, etc.), the time-dependence of the operators is understood in the Heisenberg picture, and $\langle\cdots\rangle = \mathrm{tr}[\rho_0\cdots]/Z$ is the density matrix of the initial state.[3] In the following we omit single particle indices, i.e., $G^{<}(t,t')$ and $G^{>}(t,t')$ can be viewed as an $L \times L$ matrix.

The functions $G^{<}(t,t')$ and $G^{>}(t,t')$ give the amplitude for the propagation of a hole or an additional electron in the many-body state, and therefore fully characterize what can be called the single-electron dynamics in the solid. For illustration, let us discuss how these propagators allow to distinguish between *coherent and incoherent single-particle particle propagation*: Assume we add a particle in orbital $|a\rangle$ to the many-body state $|\Psi(t)\rangle$, leading to a state $|\phi_{a,t}\rangle \equiv c^{\dagger}_{a}|\Psi(t)\rangle$, and we let this state evolve until $t' > t$ into a new state $|\phi_{a,t;t'}\rangle$. If $|\phi_{a,t;t'}\rangle$ can still be represented as the addition of a single particle to the state $|\Psi(t')\rangle$, i.e., if $|\phi_{a,t;t'}\rangle = \tilde{c}^{\dagger}|\Psi(t')\rangle$ where $\tilde{c}^{\dagger}$ is any linear combination of the $c^{\dagger}_{a}$, which creates a particle in an orbital $|\tilde{c}\rangle$, we can say that the additional particle has evolved from $|a\rangle$ to $|\tilde{c}\rangle$ independent of the

[3] The convention how to choose the $\pm i$ factors and the ordering of the time-arguments in (3.11) and (3.12) will appear more natural in the contour-ordered Keldysh formalism, Sect. 3.3.3.

other particles. Mathematically, if this holds for each state a, $G(t', t)$ can be written in terms of a unitary propagation matrix $u(t, t')$,[4]

$$G^{>,<}(t', t) = u(t', t)G^{>,<}(t, t). \tag{3.13}$$

In general, the propagation of single-particle excitations is not unitary. Deviations arise for open systems which can exchange particles with the environment, or because of interactions, which transform the single-particle excitation into excitations comprised of one particle and an arbitrary number of particle-hole pairs added to $|\Psi(t)\rangle$, such as $c^\dagger_{a_1} c^\dagger_{a_2} c_{a_3}|\Psi(t)\rangle$. The propagators therefore show to what extent a system can be described as independent quasiparticles.

For an equilibrium situation, we can separate spectral and occupation information in the fermionic propagators $G^<$ and $G^>$, analogous to the discussion for bosonic elementary excitations in the previous section. For clarity, we formulate the following paragraph indicating only a momentum index $\boldsymbol{k}$ to the Green's functions (in a translationally invariant system, $G_{\boldsymbol{k}} \equiv G_{\boldsymbol{k},\boldsymbol{k}}$ is diagonal in momentum). Spin and orbital indices can be easily added. Translational invariance in time implies that the propagators (3.11) and (3.12) depend only on the time-difference, and we can introduce the Fourier transform $G(\omega) = \int dt\, e^{i\omega t} G(t, 0)$. Similar to the derivation of the fluctuation-dissipation theorem in Sect. 3.2.1, i.e., using an expansion of $G^{>,<}_{\boldsymbol{k}}(t)$ in eigenstates of H, we can obtain the relations

$$\frac{G^<_{\boldsymbol{k}}(\omega)}{2\pi i} = A_{\boldsymbol{k}}(\omega) f(\omega) \equiv N^<_{\boldsymbol{k}}(\omega) \tag{3.14}$$

$$\frac{G^>_{\boldsymbol{k}}(\omega)}{-2\pi i} = A_{\boldsymbol{k}}(\omega)[1 - f(\omega)] \equiv N^>_{\boldsymbol{k}}(\omega), \tag{3.15}$$

where we introduced the (many-body) spectral function $A_{\boldsymbol{k}}(\omega)$, which is defined in terms of the so-called retarded Green's function as

$$A_{\boldsymbol{k}}(\omega) = -\frac{1}{\pi}\mathrm{Im} \int dt e^{i(\omega+i0)t} G^R_{\boldsymbol{k}}(t), \tag{3.16}$$

$$G^R_{\boldsymbol{k}}(t - t') = \theta(t - t')(G^>_{\boldsymbol{k}}(t - t') - G^<_{\boldsymbol{k}}(t - t')). \tag{3.17}$$

The relation between G^R and $G^{>,<}$ is the same as between response (3.3) and fluctuations (3.4) and (3.5) in the bosonic case. Unitary propagation (3.13) of the excitation would imply $G^{>,<}_{\boldsymbol{k}}(t, t') = e^{i\epsilon_{\boldsymbol{k}}(t-t')} G^{>,<}_{\boldsymbol{k}}(t', t')$ with some $\epsilon_{\boldsymbol{k}}$, and Fourier transform therefore gives a spectral function $A_{\boldsymbol{k}}(\omega) = \delta(\omega - \epsilon_{\boldsymbol{k}})$ with a well-defined energy-momentum relation. The spectrum defines the level-structure of the single-particle excitations. For a many-body system, such as an interacting metal, the perfect delta-peaks will broaden. As long as they remain centred around a well-defined dispersion $\epsilon_{\boldsymbol{k}}$, we can speak of quasiparticles with a finite lifetime. In non-equilibrium, the description of electronic structure will be based on the two-time Green's functions (3.11) and (3.12) rather than on spectral functions $A_{\boldsymbol{k}}(\omega)$.

---

[4]In Sect. 3.3.4 we show explicitly that this holds for a general quadratic Hamiltonian.

3.2.3 Probabilistic Interpretation of Real-Time Green's Functions

To further illustrate the two-time propagators, we relate these quantities to an actual measurement. For this purpose, let us introduce the Wigner transform $G^{>,<}(t,\omega)$ [(3.9)] of the propagators (3.11) and (3.12), and the corresponding occupation functions $N_{\boldsymbol{k}}^{>,<}(t,\omega)$ [(3.14) and (3.15)]. Following the equilibrium interpretation of $N_{\boldsymbol{k}}^{<}(\omega)$ as the product of a spectral function (density of states) and an occupation function, one may assume that $N_{\boldsymbol{k}}^{<}(t,\omega)$ gives the probability distribution to remove a particle with energy $\hbar\omega$ and momentum $\boldsymbol{k}$ from the system *at time* t, in an experiment such as photoemission or tunneling microscopy. This probabilistic interpretation of $N_{\boldsymbol{k}}^{<}(t,\omega)$ is "almost true", as we will see now. One only has to take care of the energy-time uncertainty relation which forbids to specify the energy transfer of a of a process which takes place at a given moment of time. We will illustrate this using the theoretical description of an ideal time-resolved photoemission experiment.

In a time- and angle-resolved photoemission experiment, one measures the probability that an electron is emitted under the action of a short probe pulse, as a function of the photoelectron energy E and the photo-electron momentum $\boldsymbol{k}_e$. The probe pulse has a finite duration, and its delay t_p with respect to a given excitation ("time zero") is varied. In this picture, the detection process of the photo-electron has no time-resolution, but the time-resolution comes from the finite interaction time of the probe light with the sample. We will assume that the electric field of the probe light has a time-profile with some envelope function $S(t)$,

$$E(t) \sim S(t-t_p)e^{-i\Omega(t-t_p)} + h.c., \tag{3.18}$$

centered around a probe frequency Ω and a probe time t_p. To facilitate a theoretical description of the photoemission process, one typically makes the so called sudden approximation, which assumes that there is no interaction between the electrons in the solid and electron in an outgoing state $|\boldsymbol{k}_e\rangle$, which is a scattering state with asymptotic plane wave form $\langle \boldsymbol{r}|\boldsymbol{k}_e\rangle \sim e^{i\boldsymbol{k}_e\boldsymbol{r}}$. Only within this approximation can the photoemission signal be interpreted as a convolution of the spectra of the solid and of the final states. The signal can be obtained using time-dependent perturbation theory in the light-matter coupling [35, 36]. A step-by step derivation, taking into account the quantum nature of the light, can be found in [37] and will not be repeated here. The final result is

$$I = \sum_{\alpha\alpha'} p^*_{q_0,\boldsymbol{k}_e,\alpha} p_{q_0,\boldsymbol{k}_e,\alpha'} \int dt dt' e^{i(t-t')E} (-i) G^<_{\alpha\alpha'}(t+t_p, t'+t_p) S(t) S(t')^*, \tag{3.19}$$

where $E = (E_{\boldsymbol{k}} - \Omega)$ is the energy extracted from the solid, and $p_{q_0,\boldsymbol{k}_e,\alpha}$ are matrix elements between the orbital χ_α in the solid and the wave-function $\phi_{\boldsymbol{k}_e}(\boldsymbol{r})$ of the outgoing state,

$$p_{q_0,k_e,\alpha} = \int d^3r\, e^{iq_0 r} \chi_\alpha(r) \nabla \phi_{k_e}(r)^*. \tag{3.20}$$

The matrix elements are typically unknown, but to analyze the position of the peaks in the photo-emission spectrum it is often sufficient to assume some simple form. For simplicity of notation, we restrict the description of the solid to a single band, and choose α to be Bloch states $\boldsymbol{k}$. Furthermore, we assume that the matrix elements are constant up to momentum conservation $p_{q_0,k_e,k} \propto \delta_{q_0+k,k_e}$.[5] With this simplification, the signal (3.19) becomes

$$I_k(E, t_p) \propto \int dt\, dt'\, e^{iE(t-t')}(-i) G_k^<(t_p+t, t_p+t')\; S(t) S(t')^*. \tag{3.21}$$

For the purpose of these notes, this expression is sufficient to analyze the properties of the two-time Green's functions. It is important to note that while $G^<(t, t')$ is complex, $I_k(E, t_p)$ is a probability which must be always real and non-negative. This positivity can be proven directly from the Lehmann representation of the Green's function.

To further illustrate (3.21), one can consider a Gaussian probe profile $S(t) = \exp(-t^2/2\Delta t^2)$ with duration Δt, and transform (3.21) to a mixed time frequency representation. We obtain

$$I_k(E, t_p) \propto \int d\omega\, dt\; N^<(t_p + t, E + \omega)\; e^{-\frac{t^2}{\Delta t^2}} e^{-\omega^2 \Delta t^2}, \tag{3.22}$$

where $N_k^<(t, \omega)$ is the Wigner transform (3.9) of the Green's function as defined above. This equation clarifies the relation between the non-negative probability $I_k(E, t_p)$ and the Wigner transform: $N_k^<(t_p, E)$ can in general be positive and negative, but it becomes positive when averaged over a time and frequency window which satisfies the frequency-time uncertainty $\Delta t \Delta\omega > 1$, and can then be interpreted as the probability to extract a particle with energy E at time t_p, up to the energy-time uncertainty. This is similar to the Wigner phase-space density in semi-classical physics, which is not a probability distribution, but becomes positive after averaging position and momentum over a phase-space volume of order $\hbar$.

An important consequence of the above discussion is that the functions (3.14) and (3.15) and the spectral functions are positive definite in a non-equilibrium steady-state, as stated already in Sect. 3.2.2 [take $\Delta t \to \infty$ in (3.22)]. In a time-evolving state, the discussion provides the basis for a separation of spectral and occupation information which underlies the derivation of kinetic equations [38, 39]: If the evolution with average time t becomes so slow that we can assume that $A_k(t, \omega)$ is constant on a timescale Δt, spectral functions and occupied/unoccupied density are positive when course-grained over frequency scale $\propto (1/\Delta t)$. We can then define a distribution function $F_k(t, \omega) = N_k^<(t, \omega)/A_k(t, \omega)$, whose evolution is governed by a quantum kinetic equation. For many fast processes in correlated systems (see Sect. 3.6),

[5] In principle, momentum conservation holds only parallel to the surface. We can imagine a layered material, where only the parallel momentum component has to be considered.

this separation of timescales will however not work. In this case the actual result of the photoemission experiment will depend on the probe profile, and the probe pulse can be shaped in ways to optimally retrieve the information on $G(t, t')$ [37].

3.3 The Keldysh Formalism

The Keldysh formalism allows to generalize concepts of many-body theory, in particular path integrals and diagrammatic perturbation theory, to describe the time-evolution and non-equilibrium steady states of interacting quantum systems. On an abstract level, this is achieved by deforming the time-contour from imaginary time (Matsubara formalism) to real time. In the following we mainly describe the formalism for the time-evolution out of a thermal equilibrium state. This section will only briefly cover those aspects of the Keldysh formalism which are needed for the later development of non-equilibrium DMFT. For an in-depth introduction into the formalism, the reader may consider the text-books by Haug and Jauho [40], Kamenev [38], or Stefanucci and van Leeuwen [41].

3.3.1 The Time-Evolution Operator

For an isolated system with Hamiltonian $H(t)$, the solution of the Schrödinger equation $i\partial_t|\psi(t)\rangle = H(t)|\psi(t)\rangle$ with initial condition $|\psi(t_0)\rangle = |\psi_0\rangle$ can be written in terms of the unitary time-evolution operator $\mathscr{U}(t, t_0)$

$$|\psi(t)\rangle = \mathscr{U}(t, t_0)|\psi_0\rangle, \tag{3.23}$$

which is defined by the differential equation $i\partial_t\mathscr{U}(t, t') = H(t)\mathscr{U}(t, t')$ with initial condition $\mathscr{U}(t, t') = 1$. Because the Schrödinger equation is an ordinary linear differential equation whose solution at time $t > t'$ is entirely determined by the state at time t', the time-evolution over successive time-intervals can be simply enchained,

$$\mathscr{U}(t_2, t_0) = \mathscr{U}(t_2, t_1)\mathscr{U}(t_1, t_0) \quad \text{for } t_2 \geq t_1 \geq t_0. \tag{3.24}$$

We can therefore divide the time-interval $[t', t]$ in infinitesimal intervals of length δt, and represent the time-evolution operator by the product (assuming $t > t'$)

$$\mathscr{U}(t, t') \approx e^{-i\delta t H(t-\delta t)} \cdots e^{-i\delta t H(t'+2\delta t)} e^{-i\delta t H(t'+\delta t)} e^{-i\delta t H(t')}. \tag{3.25}$$

Here we have used that the Hamiltonian is roughly constant along each infinitesimal time interval, so that $\mathscr{U}(t + \delta t, t) = e^{-i\delta t H(t)} + \mathscr{O}(\delta t^2)$. Equation (3.25) can be formally written as

$$\mathscr{U}(t,t') = T_t e^{-i\int_{t'}^{t} d\bar{t} H(\bar{t})}, \tag{3.26}$$

where the *time-ordering operator* T_t brings operators at later time to the left. The adjoint of (3.25) is then simply the inverse (backward) time-evolution,

$$\mathscr{U}(t,t')^\dagger = e^{i\delta t H(t')} e^{i\delta t H(t'+\delta t)} e^{i\delta t H(t'+2\delta t)} \cdots e^{i\delta t H(t-\delta t)} = T_{\bar{t}} e^{+i\int_{t'}^{t} d\bar{t} H(\bar{t})}, \tag{3.27}$$

where $T_{\bar{t}}$ is the anti time-ordering operator which brings operators at later times to the right. Note that an equivalent way of reading (3.26) is to expand the exponential function as a Taylor series,

$$\mathscr{U}(t,t') = T_t\Big\{\sum_{n=0}^{\infty}\frac{(-i)^n}{n!}\int dt_1\cdots dt_n\, H(t_1)\cdots H(t_n)\Big\} \tag{3.28}$$

$$= \sum_{n=0}^{\infty}(-i)^n\int_{t'}^{t} dt_1\int_{t'}^{t_1} dt_2\cdots\int_{t'}^{t_{n-1}} dt_n\, H(t_1)\cdots H(t_n). \tag{3.29}$$

This form is the familiar form for the time-evolution operator which is obtained by successively iterating the integral variant of the Schrödinger equation, $\mathscr{U}(t,t') = 1 - i\int_{t'}^{t} d\bar{t}\, H(\bar{t})\mathscr{U}(\bar{t},t')$.

3.3.2 Time-Dependent Expectation Values and the Keldysh Contour

In order to describe the time-evolution of a quantum system, we aim to compute observables or correlation functions of the general form

$$\langle \mathscr{O}(t_1)\rangle = \mathrm{tr}\big[\rho_0\, \mathscr{U}(t_1,t_0)^\dagger\, \mathscr{O}\, \mathscr{U}(t_1,t_0)\big]. \tag{3.30}$$

Here ρ_0 is the density matrix which defines the state of the system at initial time t_0. Note that (3.30) describes the evolution of an isolated quantum system, without contact to environment: If $\rho_0 = \sum_n w_n|\psi_n\rangle\langle\psi_n|$ is the statistical mixture of states $|\psi_n\rangle$ with weights w_n, (3.30) can be written as

$$\langle \mathscr{O}(t)\rangle = \sum_n w_n \langle\psi_n(t)|\mathscr{O}|\psi_n(t)\rangle, \tag{3.31}$$

where $|\psi_n(t)\rangle = \mathscr{U}(t,t_0)|\psi_n\rangle$ is obtained by solving the unitary Schrödinger equation with initial condition $|\psi_n(t_0)\rangle = |\psi_n\rangle$. This clearly shows that the density matrix ρ_0 in (3.30) provides only the statistical weights of the initial condition for the dynam-

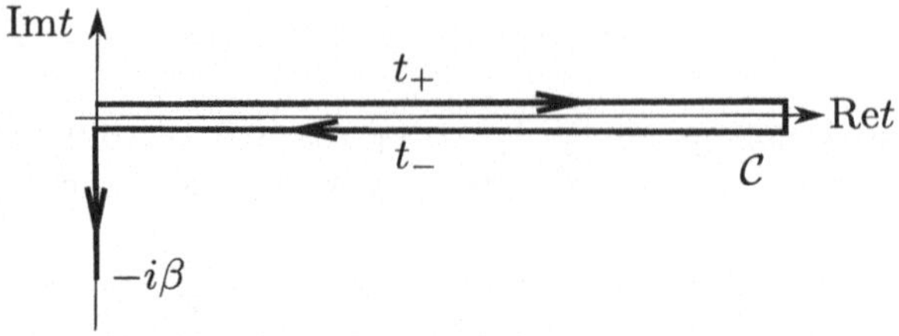

Fig. 3.1 The L-shaped Keldysh contour $\mathscr{C}$, ranging from time 0 to some maximum time t_{max}, back to time 0, and finally to $-i\beta$ on the imaginary-time branch. Times on the upper and lower real-time branch are denoted by t_+ and t_-, respectively. Note that both t_+ and t_- are purely real, and the index $\pm$ is only a book-keeping index to distinguish backward and forward time-evolution. The arrows denote the time-ordering along $\mathscr{C}$ from "earlier" to"later" contour times

ics, while the system is isolated from the environment for all times.[6] In the following, we take ρ_0 to be the Gibbs ensemble with respect to some initial Hamiltonian $H(0)$.

As a mathematical trick, the time-ordering in $\mathscr{U}(t, t_0)$ and the anti-time-ordering in $\mathscr{U}(t, t_0)^\dagger$ in (3.30) can be combined into a single time-ordering along a time contour, which first extends from t_0 in forward direction and then in backward direction.[7] Furthermore, the thermal density matrix can be written as a time-evolution operator along an imaginary time axis $[0, -i\beta]$. The three branches can be combined into a single *L-shaped contour* $\mathscr{C}$, as depicted in Fig. 3.1. The order of times, from earlier to later, is indicated by the arrows in the figure, and we will use the notation $t >_{\mathscr{C}} t'$ ($t <_{\mathscr{C}} t'$) to denote that t is later (earlier) on $\mathscr{C}$ than t'. Throughout these notes, we will denote a time argument on the upper (lower) branch by $t_\pm$ with $t \in \mathbb{R}$, respectively, and a time-argument on the vertical branch by $-i\tau$ with $\tau \in [0, \beta]$. Together with the time-contour we introduce the contour-ordering operator

$$T_{\mathscr{C}} A(t)B(t') = \begin{cases} A(t)B(t') & t >_{\mathscr{C}} t' \\ \xi B(t')A(t) & t' >_{\mathscr{C}} t \end{cases}. \tag{3.32}$$

The sign ξ is -1 if the permutation of A and B involves an odd number of permutations of fermion creation of annihilation operators, and $+1$ otherwise.[8] With this the expectation value (3.30) is written as

[6]In the discussion of thermal states, we stated that the usefulness of the Gibbs ensemble relies on the fact that the ensemble average represents the typical behavior of the majority of the individual states. By taking the Gibbs ensemble as an initial state for the time-evolution, we tacitly assume that also in the out-of-equilibrium dynamics, each time-evolved eigenstate is a typical representative of the ensemble at a given time. This is a subtle fact. One may ask on what length-scales there is a quantum analog of turbulence, in which local observables of a driven system develop large fluctuations out of infinitesimal deviations in initial states.

[7]From now on, we will take $t_0 = 0$ without loss of generality.

[8] This convention will become only important in the next section, when we introduce contour-ordered Green's functions. It assures that Bose (Fermi) operators commute (anticommute) under the time ordering $T_{\mathscr{C}} [c_\alpha^{(\dagger)}(t), c_{\alpha'}^{(\dagger)}(t')]_\mp = 0$.

$$\langle O(t)\rangle = \frac{1}{Z}\mathrm{tr}[T_{\mathscr{C}} e^{-i\int_{\mathscr{C}} d\bar{t}H(\bar{t})} O(t_+)], \tag{3.33}$$

$$Z = \mathrm{tr}[T_{\mathscr{C}} e^{-i\int_{\mathscr{C}} d\bar{t}H(\bar{t})}], \tag{3.34}$$

where we have introduced integrals along the time-integrals in the natural form

$$\int_{\mathscr{C}} dt f(t) = \int_0^{t_{\max}} dt f(t_+) - \int_0^{t_{\max}} dt f(t_-) - i\int_0^{\beta} d\tau f(-i\tau). \tag{3.35}$$

For later use, let us introduce a few other straightforward definitions for performing calculus on $\mathscr{C}$: For $a <_{\mathscr{C}} b$, $\int_{\mathscr{C},a}^{b} dt f(t)$ will denote the integral restricted to $a <_{\mathscr{C}} t <_{\mathscr{C}} b$. This implies

$$\int_{\mathscr{C},a}^{b} dt\partial_t f(t) = f(b) - f(a), \tag{3.36}$$

where ∂_t is the conventional derivative with respect to the physical time (which is the same for t_+ and t_-):

$$(\partial_t f)(t_\pm) = \frac{f((t+\epsilon)_\pm) - f(t_\pm)}{\epsilon}, \tag{3.37}$$

$$(\partial_t f)(-i\tau) = \frac{f(-i(\tau+\epsilon)) - f(-i\tau)}{-i\epsilon}. \tag{3.38}$$

Furthermore, we introduce the delta-function $\delta_{\mathscr{C}}(t, t')$ so that

$$\int_{\mathscr{C}} dt\delta_{\mathscr{C}}(t, t') f(t) = f(t'), \tag{3.39}$$

and the theta function $\theta_{\mathscr{C}}(t, t')$ which is 1 for $t >_{\mathscr{C}} t'$ and 0 otherwise.

3.3.3 Contour-Ordered Green's Functions

In this section we show that the electron and hole propagators (3.11) and (3.12) appear naturally as different components of a single contour-ordered Green's function. The latter will then provide the basis for diagrammatic perturbation theory (Sect. 3.3.7). To be specific, we consider a system of fermions or bosons with a single-particle Hilbert space spanned by a basis $\{|a\rangle\}$ (momentum, spin, orbital, etc.), and let c_a and $c_a^\dagger$ denote the corresponding annihilation and creation operators. We introduce the single-particle Green's function by the contour-ordered expectation value

$$G_{ab}(t, t') = -i\langle T_{\mathscr{C}} c_a(t) c_b^\dagger(t')\rangle_S, \tag{3.40}$$

where $\langle T_{\mathcal{C}} \cdots \rangle_S = \mathrm{tr}[T_{\mathcal{C}} e^{iS} \cdots]/\mathrm{tr}[T_{\mathcal{C}} e^{iS}]$ with the action $S = -\int_{\mathcal{C}} dt H(t)$. Before clarifying the relation of (3.40) and the electron and hole propagators, let us summarize two general and important properties of $G(t, t')$ (in fact, these properties hold for any two-point correlation function on $\mathcal{C}$):

- **Causal property**: For the formal development of the theory it is sufficient to view the Green's function as a function of the abstract contour time. However, it is important to note that the values of $G(t, t')$ with t and t' on different branches of $\mathcal{C}$ are not all independent (we omit orbital indices for clarity): *The largest real-time argument can be shifted between the upper to the lower contour branch.* For example, assume that $t' < t$, then

$$G(t_+, t'_+) = \frac{1}{Z} \mathrm{tr}[e^{-\beta H(0)} \underbrace{\mathcal{U}(0, t)\mathcal{U}(t, t_{\max})}_{\mathcal{C}_-} \underbrace{\mathcal{U}(t_{\max}, t)c\mathcal{U}(t, t')c^\dagger \mathcal{U}(t', 0)}_{\mathcal{C}_+}]$$

$$= \frac{1}{Z} \mathrm{tr}[e^{-\beta H(0)} \mathcal{U}(0, t)c \underbrace{\mathcal{U}(t, t_{\max})\mathcal{U}(t_{\max}, t)}_{=1} \mathcal{U}(t, t')c^\dagger \mathcal{U}(t', 0)] = G(t_-, t'_+).$$

  In the first line, the brackets indicate the propagation along the upper and lower branch, respectively. Because the time-evolution between t and $t_{\max}$ along the upper and lower branch cancel, the operator c can be shifted between the two branches. As we will see, this symmetry implies a causal form of the equations of motion for the Green's functions (Sect. 3.4), hence we will simply call it the *causal property* of two-time functions on $\mathcal{C}$.
- **Boundary condition**: A second important property of contour-ordered Green's functions is the boundary condition on $\mathcal{C}$,

$$G(0_+, t) = \pm G(-i\beta, t), \quad G(t, 0_+) = \pm G(t, -i\beta), \tag{3.41}$$

  where the upper (lower) sign refers to the case where c and $c^\dagger$ are Bose (Fermi) operators. These boundary-conditions are derived analogous to the argument of the previous paragraph by a suitable reordering of the operators, taking into account the cyclic property $\mathrm{tr}(ABC) = \mathrm{tr}(BCA)$ of the trace.

To clarify the meaning of the contour-ordered Green's function we now show that the information contained in $G(t, t')$ with t and t' on $\mathcal{C}$ corresponds one-to-one to the particle and hole propagators (3.11) and (3.12) for $t, t' \in (-\infty, t_{\max})$: First of all, by choosing the time-arguments on the upper and lower branch of the contour, respectively, we recover the lesser and greater Green's functions (3.11) and (3.12) for $t, t' > 0$,

$$G^<(t, t') = G(t_+, t_-), \quad G^>(t, t') = G(t_-, t_+). \tag{3.42}$$

Putting the two time-arguments of $G(t, t')$ in all possible ways on the two real-time branches of the contour would lead to four different combinations, but from the causal property of G it is clear that all these components can be reduced to $G^{<}(t,t')$ and $G^{>}(t,t')$; for example, for $t > t'$, $G(t_+, t'_+) = G(t_-, t'_+) = G^{>}(t,t')$. The vertical branch, in turn, contains the information about the particle and hole propagators at $t, t' < 0$: When both time-arguments are on the vertical branch, the forward and backward evolution cancel, and $G(-i\tau, -i\tau')$ is precisely the imaginary-time Green's function of the initial state, which is translationally invariant in time. We conveniently adopt the same parametrization that is used in the Matsubara formalism and define:

$$G^M(\tau - \tau') = -iG(-i\tau, -i\tau'). \tag{3.43}$$

To go to real-times $t < 0$, we can follow the analytical continuation from imaginary to real time, known from the Matsubara formalism [20]. For completeness, let us summarize the relevant expressions: Because of the boundary condition (3.41), $G^M(\tau)$ is periodic (antiperiodic) for Bosons (Fermions) $[G^M(\tau) = \pm G^M(\tau + \beta)]$ and can be represented by the Fourier series

$$G(\tau) = \beta \sum_n e^{-i\omega_n \tau} G(i\omega_n), \quad G(i\omega_n) = \int_0^\beta d\tau \, e^{i\omega_n \tau} G(\tau), \tag{3.44}$$

with bosonic and fermionic Matsubara frequencies $\omega_n = 2n\pi/\beta$ and $\omega_n = (2n+1)\pi/\beta$, respectively ($n \in \mathbb{Z}$). $G(i\omega_n)$ can be continued from the values $i\omega_n$ to a function $G(z)$ which is analytic in the upper/lower complex half plane. The latter is just the Fourier transform of the retarded propagator $G^R(t,t') = \theta(t-t')[G^{>}(t,t') - G^{<}(t,t')]$ and thus gives the spectral function (3.16) via

$$A(\omega) = -\frac{1}{\pi} \mathrm{Im} G(\omega + i0), \tag{3.45}$$

which in turn gives the real-time functions though the Fourier transforms (3.14) and (3.15),

$$G^{<}(t,t') = i \int d\omega \, A(\omega) f(\omega) e^{-i\omega(t-t')} \quad \text{(for Fermions)}, \tag{3.46}$$

$$G^{>}(t,t') = -i \int d\omega \, A(\omega) f(-\omega) e^{-i\omega(t-t')} \quad \text{(for Fermions)}, \tag{3.47}$$

$$G^{<}(t,t') = -i \int d\omega \, A(\omega) b(\omega) e^{-i\omega(t-t')} \quad \text{(for Bosons)}, \tag{3.48}$$

$$G^{>}(t,t') = i \int d\omega \, A(\omega) b(-\omega) e^{-i\omega(t-t')} \quad \text{(for Bosons)}. \tag{3.49}$$

These relations can be proven using a Lehmann representation analogous to Sect. 3.2.1. For the L-shaped contour, they hold for $t, t' < 0$ and allow to recover the real-time propagators of the initial equilibrium state. Likewise, one can use a Lehmann representation to show that the mixed components $G(t, -i\tau)$ can be used to retrieve the particle and hole propagators with one time-argument before and one time-argument after time zero. Note that the above discussion implies that we can freely shift the vertical branch to any $t_0 < 0$, and replace the Hamiltonian on the real-time branch $t_0 < t < 0$ by the initial Hamiltonian. We can even shift it to $t_0 = -\infty$ and thus omit it altogether, if one finds a way to impose the correct initial condition at $t_0 = -\infty$. While the L-shaped contour is most convenient to discuss time-propagation from a given state in a numerical framework, the infinite contour is suitable for analytical derivations and for the discussion of non-equilibrium steady states.

To conclude this subsection we summarize a number of definitions which are commonly used in literature to parametrize the contour-ordered Green's functions

$$\text{retarded components}:\ G^R(t,t') = \theta(t-t')[G^>(t,t') - G^<(t,t')], \tag{3.50}$$

$$\text{advanced components}:\ G^A(t,t') = \theta(t'-t)[G^<(t,t') - G^>(t,t')], \tag{3.51}$$

$$\text{Keldsyh components}:\ G^K(t,t') = G^<(t,t') + G^>(t,t'), \tag{3.52}$$

$$\text{mixed components}:\ G^{tv}(t,\tau) = G(t_\pm, -i\tau),\quad G^{vt}(\tau,t) = G(-i\tau, t_\pm). \tag{3.53}$$

The retarded component contains the spectral information, c.f. (3.16) and (3.17) and (3.46) to (3.49).

3.3.4 Noninteracting Green's Function

Equation of Motion

In the following sections we will discuss how to obtain the contour-ordered Green's functions for a given model. We start from a system of independent particles with a general quadratic Hamiltonian

$$H(t) = \sum_{a,b} c_a^\dagger h_{a,b}(t) c_b. \tag{3.54}$$

For a compact notation, we introduce L-dimensional spinors (column vectors) $\hat{\psi} = (c_1, ..., c_L)^t$, an $L \times L$ matrix $\hat{h}(t)$ in orbital space, so that $H(t) = \hat{\psi}^\dagger \hat{h}(t)\hat{\psi}$, and an analogous matrix notation for the Green's functions $\hat{G}(t,t') = -i\langle T_{\mathscr{C}} \hat{\psi}(t)\hat{\psi}^\dagger(t')\rangle$, i.e., $[\hat{G}(t,t')]_{a,a'} = G_{a,a'}(t,t')$. It is now straightforward to derive equations of motion for the Green's functions. For $t \neq t'$ we simply have

$$\partial_t \hat{G}(t,t') = -i\langle T_{\mathscr{C}} i[H(t), \hat{\psi}(t)]\hat{\psi}^\dagger(t')\rangle = -i\hat{h}(t)\hat{G}(t,t'), \tag{3.55}$$

where the first equality holds for any Hamiltonian, while the second follows from the commutator $[H, \psi] = -\hat{h}\hat{\psi}$ for the quadratic Hamiltonian (3.54). For $t = t'$, care has to be taken because of the contour-ordering of $\hat{\psi}(t)$ and $\hat{\psi}^\dagger(t')$. We have

$$\int_{\mathscr{C},t-\epsilon}^{t+\epsilon} d\bar{t}\, \partial_{\bar{t}}\, G_{a,a'}(\bar{t},t) = G_{a,a'}(t+\epsilon,t) - G_{a,a'}(t-\epsilon,t) \tag{3.56}$$

$$= -i\langle T_{\mathscr{C}}\, c_a(t)c_{a'}^{\dagger}(t)\rangle_S \pm i\langle T_{\mathscr{C}}\, c_{a'}^{\dagger}(t)c_a(t)\rangle_S = -i\langle [c_a(t), c_{a'}^{\dagger}(t)]_{\mp}\rangle_S = -i\delta_{aa'}. \tag{3.57}$$

when $t \pm \epsilon$ denotes a time which is infinitesimally later (earlier) than t on $\mathscr{C}$. Hence $\partial_t \hat{G}(t,t')$ at $t = t'$ is proportional to the delta-function (3.39). We can combine (3.55) and (3.57), leading to

$$i\partial_t \hat{G}(t,t') - \hat{h}(t)\hat{G}(t,t') = \delta_{\mathscr{C}}(t,t'), \tag{3.58}$$

$$-i\partial_{t'} \hat{G}(t,t') - \hat{G}(t,t')\hat{h}(t') = \delta_{\mathscr{C}}(t,t'). \tag{3.59}$$

(The second equation for $\partial_{t'}$ is derived in an analogous manner.)

Inverse Green's Function

Equations (3.58) and (3.59) can be written in a short form

$$G^{-1} * G = G * G^{-1} = 1, \tag{3.60}$$

if we introduce the convolution

$$[\hat{A} * \hat{B}](t,t') = \int_{\mathscr{C}} d\bar{t}\, \hat{A}(t,\bar{t})\hat{B}(\bar{t},t'), \tag{3.61}$$

and the *inverse Green's function*

$$\hat{G}^{-1}(t,t') = [i\partial_t - \hat{h}(t)]\delta_{\mathscr{C}}(t,t'). \tag{3.62}$$

The "1" in (3.60) denotes the delta-function in time and the identity matrix in orbital indices. To get from the second equation in (3.60) to (3.59) one has to peform a partial integration. From the definition of an "inverse" operator $\hat{G}^{-1}$ we would expect that it uniquely determines $\hat{G}$. While the equation of motions and $\hat{G}^{-1}$ are obtained from $\hat{G}$ in a straightforward and unique way, the inverse operation is a first-order differential equation, which requires an initial or boundary condition for a unique solution. In the present case, the boundary condition is given by (3.41). In other words, the inverse (3.62) is only well-defined in the space of two-time functions which satisfy the boundary condition (3.41). This implicit restriction of differential and integral operators on to (anti)-periodic functions on $\mathscr{C}$ will be kept throughout these notes.

Independent Particles and Unitary Evolution

It is illustrating to show from the equations of motion (3.58) and (3.59) that the propagators of independent particles correspond to a unitary time-evolution matrix, as discussed in relation to footnote 4. For the real-time components $G^<(t,t')$ and $G^>(t,t')$, the time arguments lie on different branches of the contour, so that the delta-function in (3.58) and (3.59) does not contribute, and $G^<(t,t')$ and $G^>(t,t')$ satisfy homogeneous differential equations $[i\partial_t - \hat{h}(t)]\hat{G}^{>,<}(t,t') = 0$, which are solved with a suitable initial condition. The result is[9]

$$\hat{G}^{>,<}(t,t') = \hat{u}(t,t')\hat{G}^{>,<}(t',t') \quad \text{for } t > t', \tag{3.63}$$

$$\hat{G}^{>,<}(t,t') = \hat{G}^{>,<}(t,t)\hat{u}(t',t)^\dagger \quad \text{for } t' > t, \tag{3.64}$$

where $\hat{u}(t,t') = T_t e^{-i\int_{t'}^{t}\hat{h}(s)ds}$ is the single-particle time-evolution operator. This is the basic property of free particle propagation mentioned already in Sect. 3.2.2. The equal-time propagator is just the one-particle density matrix $\rho_{a,a'}(t) = \langle c_{a'}^\dagger c_a\rangle$. For Fermions,

$$-i\hat{G}^<(t,t) = \hat{\rho}(t), \quad i\hat{G}^>(t,t) = \hat{1} - \hat{\rho}(t). \tag{3.65}$$

Using the retarded and advanced propagators (3.50) and (3.51) we therefore have

$$\hat{G}^R(t,t') = -i\theta(t-t')\hat{u}(t,t') = \hat{G}^A(t',t)^\dagger, \tag{3.66}$$

$$\hat{G}^<(t,t') = \hat{\rho}(t)\hat{G}^A(t,t') - \hat{G}^R(t,t')\hat{\rho}(t') \tag{3.67}$$

$$\hat{G}^>(t,t') = \hat{G}^R(t,t')(\hat{1} - \hat{\rho}(t')) - (\hat{1} - \hat{\rho}(t))\hat{G}^A(t,t'). \tag{3.68}$$

These equations express the separation of spectral information (retarded and advanced functions) and occupation of single particle levels. An exact separation in this form is only possible for independent particles, but it provides an important approximation for interacting systems for the derivation of kinetic equations [40].

3.3.5 *The Two-Time Self-energy*

As discussed in Sect. 3.2.2, whenever a system is not isolated or does not consist of independent particles, the propagation of single-particle excitations is not of the unitary form (3.63). The non-unitary propagation can always be written in the form

$$\hat{G}^{-1}(t,t') = [i\partial_t - \hat{h}(t)]\delta_{\mathscr{C}}(t,t') - \hat{\Sigma}(t,t'), \tag{3.69}$$

---

[9] We use (3.58) with initial condition $\hat{G}^{>,<}(t',t')$ for $t > t'$ and (3.59) with initial condition $\hat{G}^{>,<}(t,t)$ for $t' > t$.

where the self-energy $\hat{\Sigma}(t, t')$ is a contour function that takes the role of a retarded potential energy for the propagation of single-particle excitations. In the following section we will explain how the self-energy is obtained from diagrammatic perturbation theory for interacting systems, and for an open quantum system where the non-unitary propagation is due to particle exchange with a reservoir of states.

With the inverse $\hat{G}_0^{-1}(t, t') = [i\partial_t - \hat{h}(t)]\delta_{\mathscr{C}}(t, t')$ of the noninteracting Green's function G_0, (3.69) reads $\hat{G}^{-1} = \hat{G}_0^{-1} - \hat{\Sigma}$, which can be written as (convoluting from one side with G_0 and from the other with G)

$$\hat{G} = \hat{G}_0 + \hat{G}_0 * \hat{\Sigma} * \hat{G} = \hat{G}_0 + \hat{G} * \hat{\Sigma} * \hat{G}_0. \tag{3.70}$$

This integral equation for $\hat{G}$ in terms of $\hat{G}_0$ and $\hat{\Sigma}$ is called the Dyson equation. By iteratively inserting the right-hand side of the equation for $\hat{G}$, we obtain the equivalent series,

$$\hat{G} = \hat{G}_0 + \hat{G}_0 * \hat{\Sigma} * \hat{G}_0 + \hat{G}_0 * \hat{\Sigma} * \hat{G}_0 * \hat{\Sigma} * \hat{G}_0 + ..., \tag{3.71}$$

which will be used later.

3.3.6 *Self-energy of the Open Quantum System*

In addition to the L degrees of freedom $\hat{\psi}$, let us consider a "bath" of M single particle levels with operators $\hat{\phi} = (b_1, ..., b_M)^t$ (both $\hat{\psi}$ and $\hat{\phi}$ are fermions), and a linear coupling between the two systems, with Hamiltonian

$$H = \begin{pmatrix} \hat{\psi}^\dagger & \hat{\phi}^\dagger \end{pmatrix} \begin{pmatrix} \hat{h}(t) & \hat{\eta}(t) \\ \hat{\eta}(t)^\dagger & \hat{b}(t) \end{pmatrix} \begin{pmatrix} \hat{\psi} \\ \hat{\phi} \end{pmatrix}, \tag{3.72}$$

i.e., $\hat{\psi}^\dagger \hat{h}(t)\hat{\psi}$ and $\hat{\phi}^\dagger \hat{b}(t)\hat{\phi}$ are the Hamiltonian of the system and bath separately, and $\hat{\psi}^\dagger \hat{\eta}(t)\hat{\phi} + h.c.$ is the coupling, which simply includes hopping of the particles from the system to the bath and back. This model generically describes a system coupled to (non-interacting) metallic leads, or an impurity embedded in a metallic host. We can define the Green's functions of the system and the bath as

$$\hat{G}(t, t') = -i\langle T_{\mathscr{C}} \hat{\psi}(t)\hat{\psi}^\dagger(t')\rangle \tag{3.73}$$
$$\hat{B}(t, t') = -i\langle T_{\mathscr{C}} \hat{\phi}(t)\hat{\phi}^\dagger(t')\rangle, \tag{3.74}$$

as well as mixed Green's functions $\hat{F}(t, t') = -i\langle T_{\mathscr{C}} \hat{\phi}(t)\hat{\psi}^\dagger(t')\rangle$ and $\bar{F}(t, t') = -i\langle T_{\mathscr{C}} \hat{\psi}(t)\hat{\phi}^\dagger(t')\rangle$. Since the combined system and bath is a closed set of independent particles, we can apply (3.62) to the full Green's function to get

$$\begin{pmatrix} i\partial_t - \hat{h}(t) & -\hat{\eta}(t) \\ -\hat{\eta}(t)^\dagger & i\partial_t - \hat{b}(t) \end{pmatrix} \begin{pmatrix} \hat{G}(t,t') & \hat{F}(t,t') \\ \bar{F}(t,t') & \hat{B}(t,t') \end{pmatrix} = \delta_{\mathcal{C}}(t,t'). \tag{3.75}$$

We consider the 00- and 10-component of (3.75),

$$[i\partial_t - \hat{h}(t)]\hat{G}(t,t') - \hat{\eta}(t)\bar{F}(t,t') = \delta_{\mathcal{C}}(t,t'), \tag{3.76}$$

$$[i\partial_t - \hat{b}(t)]\bar{F}(t,t') - \hat{\eta}(t)^\dagger \hat{G}(t,t') = 0. \tag{3.77}$$

We can solve the second equation in the form

$$\bar{F}(t,t') = \int_{\mathcal{C}} dt_1\, \hat{B}_0(t,t_1)\hat{\eta}(t_1)^\dagger \hat{G}(t_1,t'), \tag{3.78}$$

where $\hat{B}_0(t,t')$ is determined by the equation of motion

$$\hat{B}_0^{-1}(t,t') = [i\partial_t - \hat{b}(t)]\delta_{\mathcal{C}}(t,t'). \tag{3.79}$$

The solution can be reinserted into (3.76), which yields the form (3.69) with a self-energy

$$\hat{\Sigma}(t,t') = \hat{\eta}(t)\hat{B}_0(t,t')\hat{\eta}(t')^\dagger. \tag{3.80}$$

By comparing (3.79) and (3.62) we can identify $\hat{B}_0(t,t')$ as the Green's function of the isolated bath. The self-energy is thus an intuitive representation of the process where a particle tunnels from the system into the bath at time t', propagates in the bath from t' to t and tunnels back into the system. This process makes the propagation of single-particle excitations in the system incoherent.

3.3.7 Diagrammatic Perturbation Theory

For an interacting system, the self-energy Σ can be expanded in a power series of the Green's function, where each term is represented as a Feynman diagram. The topology of the diagrams is the same as in the equilibrium Matsubara formalism, only internal time-integrals in the interpretation of diagrams must be taken over the contour $\mathcal{C}$ instead of over imaginary time. Here we briefly recapitulate the derivation of the perturbative expansion of Σ to confirm this fact.

Wick's Theorem

The first step of perturbation theory is to split the Hamiltonian $H(t) = H_0(t) + H_1(t)$, where $H_1(t)$ represents the interaction and is typically higher-order in the field-operators, such as a two-particle term $H_1 = \sum_{ijkl} U_{ijkl}(t)c_i^\dagger c_j^\dagger c_l c_k$, while H_0 is quadratic, as in (3.54). We then simply expand the evolution operator in a Taylor series,

$$G_{ab}(t,t') = -i\frac{1}{Z}\mathrm{tr}\big[T_{\mathscr{C}}\, e^{-i\int_{\mathscr{C}} d\bar{t} H_0(\bar{t})} e^{-i\int_{\mathscr{C}} d\bar{t} H_1(\bar{t})} c_a(t)c_b^\dagger(t')\big] \tag{3.81}$$

$$= \frac{Z_0}{Z}\sum_{n=0}^{\infty}\frac{(-i)^{n+1}}{n!}\int_{\mathscr{C}} dt_1\cdots dt_n \big\langle T_{\mathscr{C}}\, H_1(t_1)\cdots H_1(t_n)\, c_a(t)c_b^\dagger(t')\big\rangle_{S_0}, \tag{3.82}$$

where we have introduced the contour-ordered expectation value

$$\langle T_{\mathscr{C}} \cdots \rangle_{S_0} = \frac{1}{Z_0}\mathrm{tr}[T_{\mathscr{C}}\, e^{iS_0}\cdots],\quad Z_0 = \mathrm{tr}[T_{\mathscr{C}}\, e^{iS_0}] \tag{3.83}$$

with respect to the non-interacting action $S_0 = -\int_{\mathscr{C}} dt\, H_0(t)$. The integrand in (3.82) contains correlation functions with an arbitrary number of creation and annihilation operators. To simplify these terms we can use Wick's theorem, which states that n-particle contour-ordered expectation values for a *quadratic action* can be factorized into a determinant of two-point correlation functions (for fermions),

$$\langle T_{\mathscr{C}}\, c_1\cdots c_n c_{n'}^\dagger \cdots c_{1'}^\dagger\rangle_{S_0} = \det\big(M\big),\quad M_{jj'} = \langle T_{\mathscr{C}}\, c_j c_{j'}^\dagger\rangle_{S_0}, \tag{3.84}$$

where $c_j \equiv c_{a_j}(t_j)$ is a shorthand notation for time and orbital indices (for bosons we would obtain a permanent). The fact that Wick's theorem holds for contour-ordered Green's functions is the key insight which leads to the same structure of diagrams in the Keldysh and Matsubara formalism. Wick's theorem is proven most easily using path integrals, where it follows as a simple property of Gaussian integrals. An elementary proof using equations of motion is found in [41]. After the application of Wick's theorem, (3.82) becomes a product of two-particle Green's functions of the noninteracting system with Hamiltonian H_0. The remaining derivation of diagrammatic perturbation theory is a counting and re-labelling of the terms in the expansion, which does not depend on the type of the time-contour at all. We will not repeat these arguments here, but refer to the literature [20].

Self-energy for the Hubbard Model

For illustration and later reference, let us summarize the diagrams for the self-energy of the Hubbard model. The Hubbard Hamiltonian

$$H = H_0 + U\sum_j c_{j\uparrow}^\dagger c_{j\uparrow} c_{j\downarrow}^\dagger c_{j\downarrow}. \tag{3.85}$$

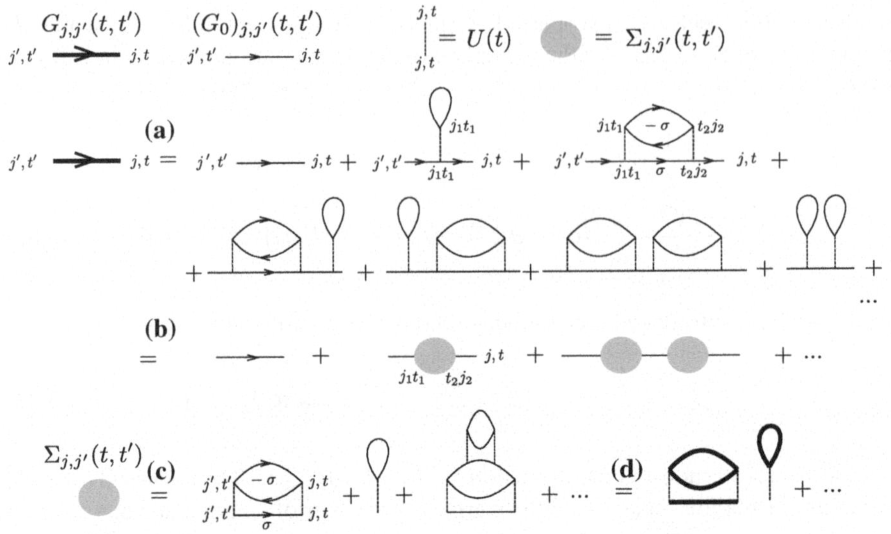

Fig. 3.2 Diagrammatic perturbation theory for the Hubbard model. The first row shows the basic elements of diagrams: Full lines between space-time points j', t' and j, t represent the interacting Green's function $G_{j,j'}(t, t')$, thin lines represent the non-interacting Green's function, and dotted lines represent the interaction, which is diagonal in space and time for the Hubbard model. The second row [(a)] shows the expansion of the Green's function, (b) to (d) show the re-summation of the expansion using the self-energy (see text)

describes electrons with spin σ which hop between sites j on a lattice (H_0 is a non-interacting tight-binding Hamiltonian), and are subject to a local interaction U. The diagrammatic representation of G and Σ is illustrated in Fig. 3.2: Green's functions $G_{jj'}(t, t')$ and noninteracting Green's function $G_{0,jj'}(t, t')$ are represented by full and thin directed lines, respectively, and the interaction $U(t)$ is denoted by a dotted line. The Green's function is given by all topologically inequivalent connected diagrams from a space-time point j', t' (particle creation) to a space-time point j, t (particle annihilation), with an arbitrary number of internal lines [(a) in Fig. 3.2]. A diagram at order n is evaluated as follows: Label all internal vertices, multiply the corresponding G_0 and U factors, multiply with $(i)^n$, multiply with -1 if there is an odd number of fermion loops, sum over the internal space indices, and integrate internal times over the contour $\mathscr{C}$. In a tadpole diagram, such as the second diagram on the right-hand side of (a), the time-argument of the creation operator must be taken infinitesimally later than the one of the annihilation operator (normal ordering). The series can then be re-summed into a Dyson equation (3.71) [(b) in Fig. 3.2] by defining the self-energy as the sum over all diagrams which are one-particle irreducible, i.e. the cannot be separated into parts by cutting a single line, as shown in (c) in Fig. 3.2. For example, when the first two terms of the expansion in (c) are inserted into the first two terms of (b), all terms of (a) are generated. The rules for Σ follow from the rules for G. To first order, the self-energy is given by ($\bar{\sigma} = \downarrow$ for $\sigma = \uparrow$ and vice versa)

$$\Sigma_\sigma^{(1)}(t,t') = U(t)n_{\bar{\sigma}}(t)\delta_{\mathscr{C}}(t,t'), \tag{3.86}$$

and the self-energy at second order is given by

$$\Sigma_\sigma^{(2)}(t,t') = U(t)U(t')G_{0\sigma}(t,t')G_{0\bar{\sigma}}(t',t)G_{0\bar{\sigma}}(t,t'). \tag{3.87}$$

Self-consistent Expansions

The diagrammatic expansion generates an expression for $\Sigma[G_0]$ as a functional of G_0. A further re-summation can be achieved by removing all terms from the expansion of Σ in which the G_0-lines itself have self-energy insertions, such as the third diagram in (c) in Fig. 3.2, and in turn replace all Green's function lines in the Σ diagrams by the full interacting self-energy G [(d) in Fig. 3.2]. The resulting expression is a functional $\Sigma[G]$ which is called the skeleton functional. A certain approximation to the skeleton functional, e.g., the truncation to a certain diagram class in combination with the Dyson (3.69) yields a closed-form non-linear integral-differential equation for G,

$$G_0^{-1} * G - \Sigma[G] * G = 1. \tag{3.88}$$

The skeleton expansion to given order generates already an infinite subset of the diagrams of the bare expansion $\Sigma[G_0]$. Contrary to a naive expectation, this does not always lead to quantitatively better results for the spectral properties in equilibrium. Important examples are the second order perturbation theory in DMFT (IPT, see Sect. 3.5.3). However, one can show that the self-consistent skeleton expansion $\Sigma[G]$ satisfies conservation laws for energy and particle number, while an approximation to $\Sigma[G_0]$ does not [42]. This is of course of great importance for the description of non-equilibrium dynamics. For example, if a theory does not respect energy conservation, it becomes meaningless to ask whether an isolated system thermalizes, i.e., whether its state at long times becomes identical to the state of a system in thermal equilibrium *at the same energy*.

3.4 The Dyson Equation

For a given electronic self-energy, the determination of the Green's function requires the solution of the Dyson equation (3.70). In real-time, the Dyson equation corresponds to a non-Markovian equation of motion for the Green's function, with the self-energy as memory kernel. In this section we explain how this equation can be solved numerically, and discuss the limit of a non-equilibrium steady state.

3.4.1 Langreth Rules

In order to work with contour-functions in practice, we have to find a suitable parametrization of the two-time contour functions in terms of objects which depend on real or imaginary time. Because of the causal property of Green's functions (Sect. 3.3.3) there are several equivalent possibilities: The real-time part of G can be parametrized either in terms of $G^>$ and $G^<$, or in terms of G^R, G^A and either of the functions G^K, $G^<$ or $G^>$. The mixed and imaginary time components are parametrized in terms of $G^{\rceil}$ and G^M [(3.50) to (3.53)]. Because the Dyson equation contains the convolution of two functions along $\mathscr{C}$, the main technicality consists in finding the real- and imaginary time components of the convolution $A * B$ of two contour functions in terms of the real- and imaginary time functions of the factors A and B. These relations are the so-called *Langreth rules*.

For the derivation of these rules, we first focus only on the real-time branches of $\mathscr{C}$. The part of the convolution integral which involves imaginary times can easily be added later on. With the two branches, one can start from a parametrization of G in terms of a 2×2 matrix,[10]

$$\hat{G}(t,t') \equiv \begin{pmatrix} G(t_+,t'_+) & G(t_+,t'_-) \\ G(t_-,t'_+) & G(t_-,t'_-) \end{pmatrix}, \quad \text{for } t,t' \in \mathbb{R}. \tag{3.89}$$

In terms of the 2×2 representation, the convolution

$$C(t_\alpha, t'_{\alpha'}) = \int_0^t d\bar{t}\big[A(t_\alpha,\bar{t}_+)B(\bar{t}_+,t'_{\alpha'}) - A(t_\alpha,\bar{t}_-)B(\bar{t}_-,t'_{\alpha'})\big], \quad \alpha,\alpha' \in \pm \tag{3.90}$$

of two contour functions $A(t,t')$ and $B(t,t')$ can be written in the form ($\hat{\tau}_3 = \mathrm{diag}(1,-1)$ is the third Pauli matrix)

$$\hat{C}(t,t') = \int_0^t d\bar{t}\hat{A}(t,\bar{t})\hat{\tau}_3\hat{B}(\bar{t},t'). \tag{3.91}$$

Because of the causal property of Green's functions, the individual entries of the matrix $\hat{G}$ are related, so that the representation (3.89) is over-complete. This over-completeness can be removed by the Keldysh rotation, which is defined by

$$\underline{G}(t,t') \equiv \hat{L}\hat{\tau}_3\hat{G}(t,t')\hat{L}^\dagger, \quad \text{where } \hat{L} = \frac{1}{\sqrt{2}}\begin{pmatrix} +1 & -1 \\ +1 & +1 \end{pmatrix}. \tag{3.92}$$

From the rotation we get (with a notation $G_{ab} = G(t_a, t'_b)$ for $a, b = \pm$)

[10] In this section we use the hat to indicate the 2×2 matrix.

$$\underline{G}(t,t') = \frac{1}{2}\begin{pmatrix} (G_{++} - G_{+-} + G_{-+} - G_{--}) & (G_{++} + G_{+-} + G_{-+} + G_{--}) \\ (G_{++} - G_{+-} - G_{-+} + G_{--}) & (G_{++} + G_{+-} - G_{-+} - G_{--}) \end{pmatrix}$$
$$= \begin{pmatrix} G^R(t,t') & G^K(t,t') \\ 0 & G^A(t,t') \end{pmatrix}. \tag{3.93}$$

The second equality follows from the causality property. For example, in the 00-component we can shift the first time argument between the upper and lower contour for $t > t'$, which gives $\frac{1}{2}(G_{-+} - G_{+-} + G_{-+} - G_{+-}) = G^> - G^<$, while for $t < t'$ we can shift the second argument, which gives $\frac{1}{2}(G_{+-} - G_{+-} + G_{-+} - G_{-+}) = 0$. In combination, $\frac{1}{2}(G_{++} - G_{+-} + G_{-+} - G_{--}) = \theta(t-t')(G^> - G^<) = G^R(t,t')$. The other components follow analogously. We then perform the Keldysh rotation in (3.91),

$$\underline{C}(t,t') \equiv \hat{L}\hat{\tau}_3\hat{C}(t,t')\hat{L}^\dagger = \int_0^t d\bar{t}\hat{L}\hat{\tau}_3\hat{A}(t,\bar{t})\hat{L}^\dagger\hat{L}\hat{\tau}_3\hat{B}(\bar{t},t')\hat{L}^\dagger \tag{3.94}$$

$$= \int_0^t d\bar{t}\underline{A}(t,\bar{t})\underline{B}(\bar{t},t'), \tag{3.95}$$

i.e., the convolution along $\mathscr{C}$ becomes a simple convolution in real-time with an additional 2×2 matrix structure. Comparison of the matrix entries on both sides this equations,

$$\begin{pmatrix} C^R(t,t') & C^K(t,t') \\ 0 & C^A(t,t') \end{pmatrix} = \int_0^t d\bar{t} \begin{pmatrix} A^R(t,\bar{t}) & A^K(t,\bar{t}) \\ 0 & A^A(t,\bar{t}) \end{pmatrix} \begin{pmatrix} B^R(\bar{t},t') & B^K(\bar{t},t') \\ 0 & B^A(\bar{t},t') \end{pmatrix}, \tag{3.96}$$

gives the Langeth rules

$$C^{R,A}(t,t') = \int_{t'}^t d\bar{t}\, A^{R,A}(t,\bar{t})B^{R,A}(\bar{t},t') \tag{3.97}$$

$$C^K(t,t') = \int_0^t d\bar{t}A^R(t,\bar{t})B^K(\bar{t},t') + \int_0^{t'} d\bar{t}A^K(t,\bar{t})B^A(\bar{t},t'). \tag{3.98}$$

In these equations we have already taken into account that retarded components for $t < t'$ and advanced components for $t > t'$ vanish, and have restricted the time-arguments accordingly. This has an important consequence: In order to determine $C^{R,A,K}(t,t')$ in the domain $t,t' \leq t_{max}$, also A and B have to be known only for $t,t' \leq t_{max}$. This is the basis for the time-stepping algorithm explained in the next section. The Langreth rules on the L-shaped contour are a straightforward extension of (3.97) and (3.98), and are obtained by adding the integrals along the imaginary branch (see, e.g., [29]). We have

$$C^{R,A} = A^{R,A} \cdot B^{R,A}, \tag{3.99}$$

$$C^{>,<} = A^R \cdot B^{>,<} + A^{>,<} \cdot B^A + A^{tv} \cdot B^{vt}, \tag{3.100}$$

$$C^{tv} = A^R \cdot B^{tv} + A^{tv} \cdot B^M, \tag{3.101}$$

$$C^M = A^M \cdot B^M, \tag{3.102}$$

where $\cdot$ is the convolution along real or imaginary time, depending on the respective time-argument.

3.4.2 Kadanoff-Baym Equations

In this subsection we discuss the numerical solution of the Dyson equation in real time. We consider the standard form

$$(i\partial_t - h(t))G(t,t') - \int_{\mathscr{C}} d\bar{t}\, \Sigma(t,\bar{t})G(\bar{t},t') = \delta_{\mathscr{C}}(t,t'), \tag{3.103}$$

which is the explicit expression for $G^{-1} * G = 1$, using (3.69). On the L-shaped contour, this equation has a unique solution when the boundary condition (3.41) is applied. To solve the equation we could take an equidistant time grid, with N time slices on the upper and lower real-time branch, and M imaginary time slices. The operator $(i\partial_t - h(t))\delta_{\mathscr{C}}(t,t') - \Sigma(t,t')$ can then be written as a $(2N + M + 1)$-dimensional matrix, and the solution for G becomes a matrix inversion [43]. This scheme of solving the whole equation on $\mathscr{C}$ all at once, however, does not exploit the possibility to transform (3.103) into a causal time propagation scheme: If the solution of (3.103) is known for real times $t, t' \leq n\Delta t$, it can be extended to the domain $t, t' \leq (n+1)\Delta t$ without modification of the previous times. The solution of (3.103) can therefore be formulated as a time-propagation scheme, which is extremely useful if Σ itself is a functional of the Green's function, e.g., in the form of a perturbative expression $\Sigma[G]$. (In general, any functional expression $\Sigma[G]$ should be causal, such that $\Sigma(t,t')$ in the domain $t, t' \leq n\Delta t$ can be determined from G in the same domain $t, t' \leq n\Delta t$.)

To obtain the explicit causal time-propagation scheme, we parametrize G in terms of G^M, G^R, G^{tv}, and $G^<$, following the implementation in [44]. With the Langreth rules (3.99)–(3.102), and a corresponding representation of the derivative (3.37), we obtain four coupled integral equations,

$$[-\partial_\tau - h(0^-)]G^M(\tau) - \int_0^\beta d\bar{\tau}\, \Sigma^M(\tau - \bar{\tau})G^M(\bar{\tau}) = \delta(\tau), \tag{3.104}$$

$$[i\partial_t - h(t)]G^R(t,t') - \int_{t'}^{t} d\bar{t}\, \Sigma^R(t,\bar{t})G^R(\bar{t},t') = \delta(t-t'), \tag{3.105}$$

$$[i\partial_t - h(t)]G^{tv}(t,\tau') - \int_0^t d\bar{t}\, \Sigma^R(t,\bar{t})G^{tv}(\bar{t},\tau') = Q^{tv}(t,t'), \tag{3.106}$$

$$[i\partial_t - h(t)]G^{<}(t,t') - \int_0^t d\bar{t}\, \Sigma^R(t,\bar{t})G^{<}(\bar{t},t') = Q^{<}(t,t'), \tag{3.107}$$

with

$$Q^{tv}(t,t'), = \int_0^\beta d\bar{\tau}\, \Sigma^{tv}(t,\bar{\tau})G^M(\bar{\tau},\tau'), \tag{3.108}$$

$$Q^{<}(t,t') = \int_0^{t'} d\bar{t}\, \Sigma^{<}(t,\bar{t})G^A(\bar{t},t') - i\int_0^\beta d\bar{\tau}\, \Sigma^{tv}(t,\bar{\tau})G^{vt}(\bar{\tau},t'). \tag{3.109}$$

Here, the integral limits take into account that retarded functions vanish for $t < t'$. These integral equations for the real and imaginary time Green's functions are known as Kadanoff-Baym equations. An equivalent set of equations can be obtained from the conjugate equation $G * G^{-1} = 1$.

To see how these Kadanoff-Baym equations represent the above mentioned time-propagation scheme, one may first note that (3.104) for G^M is decoupled from the other equations, and can be solved with a boundary condition $G^M(\tau) = \pm G^M(\tau + \beta)$ by the usual Fourier representation (3.44). Its solution is the Green's function of the initial equilibrium state,

$$G^M(i\omega_n) = [i\omega_n - h(0^-) - \Sigma^M(i\omega_n)]^{-1}. \tag{3.110}$$

To see the causal structure of the remaining real-time equations, consider first the retarded equation (3.105). With the parametrization $y(s) = G^R(t'+s,t')$, (3.105) can be written in the standard form of a Volterra integral-differential equation of second kind,

$$\frac{d}{ds}y(s) = q(s) + p(s)y(s) + \int_0^s d\bar{s}\, k(s,\bar{s})y(\bar{s}), \tag{3.111}$$

taking $q(s) = 0$, $p(s) = -ih(s)$, $k(s,\bar{s}) = -i\Sigma^R(t'+s,t'+\bar{s})$. Stable and accurate high-order algorithms for the solution of Volterra equations can be found in the literature [45]. In the simplest case, one can work with an equidistant grid $s \in \{m\Delta t\}$, represent the integral with a Trapez rule and the differential term by a finite difference formula (using the notation $y_m = y(m\Delta t)$),

$$\frac{y_{n+1} - y_n}{\Delta t} = q_{n+1} + p_{n+1}y_{n+1} + \Delta t\Big(\tfrac{1}{2}y_{n+1} + \sum_{m=1}^{n} k_{n,m}\, y_m + \tfrac{1}{2}y_0\Big). \tag{3.112}$$

Starting from an initial value y_0 (here, $y(0) = G^R(t',t') = -i$), this equation provides an implicit equation of y_{n+1} in terms of y_m for $m \le n$, which can be successively applied for $n = 1, 2, 3, \ldots$ to determine the full function y. In practice higher order

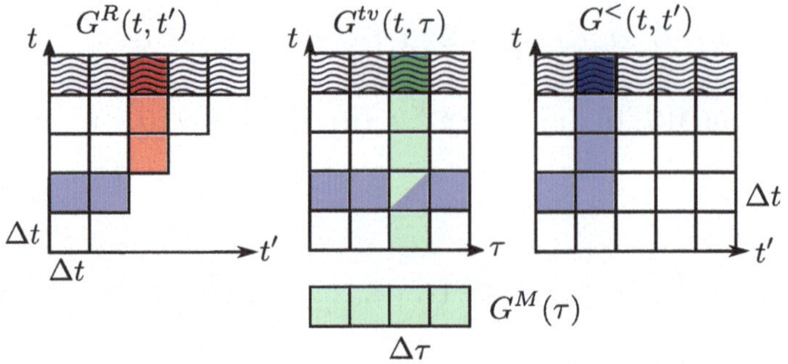

Fig. 3.3 Time-structure of the Kadanoff-Baym equations: The hatched region indicates a time-slice $T_n \equiv \{G^R(n\Delta t, m\Delta t), G^{tv}(n\Delta t, \tau), G^<(n\Delta t, m\Delta t) | 0 \leq \tau \leq \beta, m \leq n\}$. In addition, the light red region indicates the values of G which are needed for the computation of the dark-red element of G^R on T_n, the light green region indicates the values of G which are needed for the computation of the dark green element of G^{tv} on T_n, and the light blue region indicates the values of G which are needed for the computation of the dark blue element of G^R on T_n. Hermitian symmetries are used to relate G^A with G^R, and $G^<(t,t')$ with $G^<(t',t)$

accurate approximations should be used for derivative and integral [44, 45], but the basic form of the equation remains the same.

Equations (3.105)–(3.107) can all be reduced in the form (3.111) (the equations for G^{tv} and $G^<$ have a nonzero kernel $q(s)$). Figure 3.3 illustrates which elements of G enter the discrete Volterra (3.112) for the determination of G on one "time-slice" $\{G^R(n\Delta t, m\Delta t), G^{tv}(n\Delta t, \tau), G^<(n\Delta t, m\Delta t) | 0 \leq \tau \leq \beta, m \leq n\}$, and thus shows the possibility for a successive time-evolution by proceeding from one time-step to the next. For M time-slices, the required computational resources scale like $\mathcal{O}(M^2)$ for memory and $\mathcal{O}(M^3)$ for CPU time. In practice the memory is a limiting factor, in particular when Green's functions carry many orbital indices.

Finally, let us remark that the memory integral $\Sigma * G$, which makes the difference to a unitary evolution of single-particle excitations, corresponds to a generalized scattering term of a Boltzmann equation. In fact, there are several approaches to derive a Boltzmann equation from the full Kadanoff-Baym [38–40]. In one way or another, all derivations rely on two approximations, which is (i), a separation of timescales between the evolution of the spectral function and the occupation function, and (ii), the assumption of a well-defined quasi-particle form of the spectral function. In combination, one obtains an equation of motion for the distribution function alone. A rather transparent derivation makes used of the so-called generalized Kadanoff-Baym ansatz (GKBA) [40, 46], to obtain an equation of motion for the one-particle density matrix (3.65). The separation between this occupation function and the spectrum is achieved by simply using the effective unitary propagation (3.67) and (3.68) as an ansatz for the Green's function within a given functional $\Sigma[G]$. Furthermore, the quasiparticle approximation in this context corresponds to a suitable approximation of the retarded and advanced propagators. In the simplest form, one evaluates G^R and G^A with an effective mean-field Hamiltonian, which corresponds to ideal quasiparticles. In combination, the Kadanoff-Baym equation with a two-time memory integral $G * \Sigma[G^<, G^>]$ for the propagators $G^<$ and $G^>$ can be reduced to a differential equation for the density matrix (3.65) with only a single-time memory integral $I[\rho]$. Such a kinetic equation seems to be ideal to study the evolution of non-trivial condensed matter phases in the weak-coupling limit, such as superconductors or charge-density waves: For such systems the spectrum and the equilibrium state are often well described within mean-field theory, while a *time-dependent* mean-field is qualitatively wrong, as it does not include scattering processes which lead to thermalization. The latter would be added by the memory integral obtained from the kinetic equation. Interestingly, in spite of the apparent simplicity of the GKBA formalism, it is still a question of active research to what extent it can be used to study the evolution of non-trivial condensed matter phases. In these notes we focus on strongly correlated systems where a separation of spectrum and occupation is not possible anyway, and we therefore refer to the literature for a more detailed description of kinetic equations [40].

3.4.3 Steady-State Formalism

An important application of the Keldysh formalism is to study non-equilibrium steady states. In a non-equilibrium steady state, we assume that the system is simultaneously subject to a time-independent perturbation and coupled to external reservoirs, such as a system in contact to external leads with a voltage bias. If we assume that all transients after the initial switch-on of the perturbation decay with time, the system will eventually relax into a steady state in which energy (particle) absorption from the perturbation is balanced by dissipation into the bath. In this state, all correlations functions are translationally invariant in time, but the system is not in a thermal equilibrium state. The latter is indicated by a breaking of the fluctuation dissipation theorem, i.e., we have $G^<(\omega) = 2\pi i A(\omega) F(\omega)$ and $G^>(\omega) = -2\pi i A(\omega)(1 - F(\omega))$ like in (3.14) and (3.15) with a positive definite spectral function $A(\omega)$ and occupation function $F(\omega)$, but F is not given by the Fermi function. In a bulk system, such non-equilibrium steady states can even show current-induced phase transitions [47].

In the steady state, the solution of the Kadanoff-Baym equations is simplified because of time-translational invariance, which allows to define the Fourier transform. We have

$$G^R(\omega + i0) = [\omega + i0 - h - \Sigma^R(\omega + i0)]^{-1} = G^A(\omega - i0)^\dagger \tag{3.113}$$

for the retarded functions. The equation for the Keldysh component becomes

$$[\omega + i0 - h - \Sigma^R(\omega + i0)] G^K(\omega) = \Sigma^K(\omega) G^A(\omega), \tag{3.114}$$

which can be solved using (3.113),

$$G^K(\omega) = G^R(\omega)\Sigma^K(\omega)G^A(\omega). \tag{3.115}$$

In practice, we solve (3.113) and (3.115) together with a given approximation $\Sigma[G]$ for the self-energy, such as (3.87). One can often solve this equation iteratively: Starting from some a given guess for $\Sigma^R(\omega + i0)$ and $\Sigma^K(\omega)$, we compute $G^K(\omega)$, $G^R(\omega + i0)$, and the spectral function $A(\omega)$. A Fourier transform gives $G^R(t) = -i\theta(t)\int d\omega A(\omega)e^{-i\omega t}$ and $G^K(t) = \int \frac{d\omega}{2\pi} G^K(\omega)e^{-i\omega t}$. These Green's functions are then in turn used to calculate $\Sigma^R(t)$ and $\Sigma^K(t)$, which are transformed back to $\Sigma^R(\omega + i0)$ and $\Sigma^K(\omega)$, and the procedure is iterated to convergence. In many situations the functions $A(\omega)$, $\Sigma^K(\omega)$, $G^K(\omega)$ are well localized in frequency, i.e., the frequency grid can be kept finite in spite of the slowly decaying $1/\omega$ tail in $G^R(\omega)$.

Note that (3.115) seems to be ill-defined for $\Sigma^K = 0$, i.e., for a noninteracting system which is isolated from the environment. In fact, one can see that for $\Sigma^K = 0$ (3.114) is solved by any ansatz of the form

$$G_0^K(\omega) = G^R(\omega)F(\omega) - F(\omega)G^A(\omega), \tag{3.116}$$

with an arbitrary (hermitian) distribution function $F(\omega)$. This freedom should be expected on physical grounds, because for an isolated noninteracting system any distribution of the single-particle levels provides a possible initial state. By extending the real-time part of the contour to $(-\infty, \infty)$ in the steady-state formalism, we have removed the memory on the initial condition and left the steady state undetermined for $\Sigma^K = 0$. The distribution function of the noninteracting system must then be provided as as external condition. Mathematically, one way of fixing the distribution function in the solution (3.116) of (3.114) is to set $\Sigma^K(\omega) = -i\eta F(\omega)$ and let $\eta \to 0^+$. With the definition of a bath self-energy (3.80), this is equivalent of attaching a bath to the system with a flat density of states, a given distribution $F(\omega)$, and infinitesimal coupling $\sqrt{\eta}$. In the long time limit the bath determines the distribution function of the system, irrespective of the coupling strength, even though for infinitesimal coupling the spectral properties of the system are not affected.

3.5 Nonequilibrium Dynamical-Mean-Field Theory

3.5.1 *The Dynamical Mean-Field Formalism*

DMFT and its extensions nowadays present some of the most versatile approaches to describe correlated systems, and are routinely used in combination with ab-initio methods to obtain the electronic structure of complex materials [48]. Rather than being perturbative in the interaction, DMFT approximates only the spacial correlations in a mean-field manner, but accurately treats local temporal fluctuations. This fact, which becomes exact in the limit of infinite coordination number [49], has been essential to understand systems in which atomic correlations compete with the itinerant behavior of electrons. There are numerous introductory texts on DMFT, such as the lecture notes by A. Georges [50], or [51]. Like any many-body formalism, the formulation of DMFT within the Keldysh and the Matsubara framework differs only by the choice of the time contour [29, 52, 53], and we therefore explain the formalism immediately in real time.

To keep the notation simple, we present the DMFT formalism for a single-band Hubbard model,

$$H = \sum_{\langle ij \rangle, \Sigma} v_{ij}\, c_{i\sigma}^{\dagger} c_{j\sigma} + \sum_{i} h_{\text{loc,i}}, \qquad h_{\text{loc,i}} = U c_{i\uparrow}^{\dagger} c_{i\uparrow} c_{i\downarrow}^{\dagger} c_{i\downarrow}. \tag{3.117}$$

Here, $c_{i\sigma}^{\dagger}$ ($c_{i\sigma}$) create (annihilate) an electron with spin σ at site i of a crystal lattice, v_{ij} is the hopping matrix element, and electrons interact via a local Coulomb interaction U. DMFT can be formulated for Hamiltonians of more complex form, which involve a coupling to local oscillators (optical phonons), or more than one orbital per site, as long as $h_{\text{loc,i}}$ remains local. The starting point for the derivation of DMFT has been the limit of infinite dimensions [49]. In order to have a meaningful description of the physics in this limit, the hopping matrix elements must be rescaled such that the average kinetic energy remains finite, and the physically relevant competition between kinetic and interaction energy is preserved. For a d-dimensional cubic lattice with nearest neighbor hopping, this leads to

$$v = v_* / \sqrt{2d}, \tag{3.118}$$

where v_* is kept constant as $d \to \infty$. A consequence of the infinite dimensional limit is the locality of the self-energy in space [54, 55],

$$\Sigma_{ij}(t, t') = \delta_{ij}\, \Sigma_{ii}(t, t'), \tag{3.119}$$

(the dependence on Σ is omitted for notational simplicity). The proof for this fact relies only counting powers of d in the diagrams, and because the topology of the diagrams is the same in the Keldysh and Matsubara formalism, (3.119) holds also in the Keldysh formalism. The contour-ordered lattice Green's functions $G_{ij}(t, t') = -i \langle T_{\mathscr{C}}\, c_{i\sigma}(t) c_{j\sigma}^{\dagger}(t') \rangle$ can then be obtained from a Dyson equation [(3.69)]

$$(G^{-1})_{ij}(t, t') = (G_0^{-1})_{ij}(t, t') - \delta_{ij}\, \Sigma_{ii}(t, t'), \tag{3.120}$$

where $(G_0^{-1})_{ij}(t, t') = \big[\delta_{ij}(i\partial_t + \mu) - v_{ij}(t)\big] \delta_{\mathscr{C}}(t, t')$ is the inverse of the noninteracting lattice Green's function [(3.62)]. To obtain the local self-energy, one must however still sum local diagrams to infinite order, which is facilitated by the following trick:

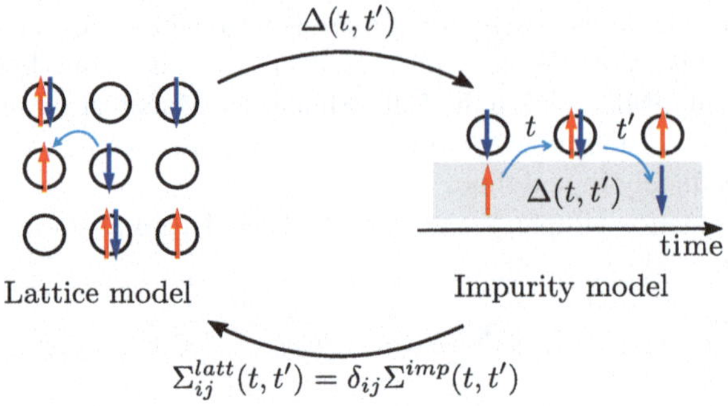

Fig. 3.4 Schematic representation of the mapping from a lattice model with local self-energy to an impurity model with hybidization function $\Delta(t, t')$. The time-dependent exchange of particles of the impurity sites with a bath represents the time-dependent fluctuations of the particle number on a site in the lattice

One can note that the locality of Σ holds also for the skeleton functional (Sect. 3.3.7): In the limit of large coordination number, the functional $\Sigma_{ii}^{skel}[G]$ can simply be restricted to the sum of all diagrams which contain the local Green's function only. The latter can be used to obtain the self-energy from an *auxiliary impurity problem* (Fig. 3.4). Consider, for example, an Anderson impurity Hamiltonian of the form

$$H_{imp} = U c_\uparrow^\dagger c_\uparrow c_\downarrow^\dagger c_\downarrow + \epsilon_f \sum_\sigma c_\sigma^\dagger c_\sigma + \sum_{p\sigma} (V_p c_\sigma^\dagger a_{p\sigma} + h.c.) + \sum_{p\sigma} \epsilon_p a_{p\sigma}^\dagger a_{p\sigma}, \tag{3.121}$$

where one interacting site is coupled to noninteracting bath orbitals p. The functional dependence of the skeleton diagrams $\Sigma_{imp}[G]$ on the Green's function does not depend on the form of the noninteracting part of the Hamiltonian, so that the diagrammatic series $\Sigma_{imp}[G]$ contains precisely the same terms as the restriction of the lattice functional $\Sigma_{ii}^{skel}[G]$ to the contributions from the local Green's function G_{ii}, as long as the interaction on the impurity is the same as the lattice model. Hence the solution of the impurity Hamiltionian (3.121) for a given Green's function G_{imp} and self energy Σ_{imp} can be seen as a device to implicitly sum the local skeleton series of the lattice model,

$$\Sigma_{ii}^{skel}[G_{imp}] = \Sigma_{imp}. \tag{3.122}$$

This fact was first noted for the Falikov-Kimball model [56]. For the Hubbard model, the auxiliary impurity formulation was developed by Georges and Kotliar [57]. The self-energy at site j can thus be obtained from a local model with the most general quadratic action

$$S_{imp,j} = -i \int_{\mathscr{C}} dt\, h_{\mathrm{loc,j}}(t) - i \sum_{\sigma} \int_{\mathscr{C}} dtdt'\, c_{\sigma}^{\dagger}(t) \Delta_j(t,t') c_{\sigma}(t'), \tag{3.123}$$

with an auxiliary hybridization function $\Delta_j(t,t')$.[11] The impurity Dyson equation fixes a relation between the impurity Green's function

$$G_{imp,j}(t,t') = -i \langle T_{\mathscr{C}} c(t) c^{\dagger}(t') \rangle_{S_{imp,j}} \tag{3.124}$$

and the impurity self-energy

$$G_{imp,j}^{-1}(t,t') = (i\partial_t + \mu)\delta_{\mathscr{C}}(t,t') - \Sigma_{imp,j}(t,t') - \Delta_j(t,t'). \tag{3.125}$$

Finally the equations are closed when the bath $\Delta(t,t')$ is determined self-consistently in such a way that both the local lattice self-energy Σ_{jj} and the local lattice Green's function G_{jj} equal the corresponding impurity quantities,

$$G_{imp,j}(t,t') \stackrel{!}{=} G_{jj}(t,t'), \quad \Sigma_{imp,j}(t,t') \stackrel{!}{=} \Sigma_{jj}(t,t'). \tag{3.126}$$

Equations (3.120), (3.123)–(3.126) provide the closed set of equations for non-equilibrium DMFT.

An equivalent and intuitive way to derive the DMFT equations is the cavity method [51]: In a lattice model with a local interaction, the degrees of freedom at different sites are only coupled bi-linearly. One can thus focus on a given site j and integrate out the degrees of freedom at all lattice sites. Keeping only the quadratic part of the resulting effective action (which remains nonzero in the infinite dimensional limit) leads to DMFT. It is illustrating to perform this cavity construction to lowest order: One can expand

$$Z = \mathrm{tr}_j \mathrm{tr}_{l\neq j} \big[T_{\mathscr{C}}\, e^{iS_j[c_j, c_j^{\dagger}]} e^{iS_{j\neq l}[c_{j\neq l}, c_{j\neq l}^{\dagger}]} e^{-i\int_{\mathscr{C}} dt \sum_{l,\sigma} v_{jl} c_{j\sigma}^{\dagger}(t) c_{l\sigma}(t)} \big] \tag{3.127}$$

to second order in the inter-site-coupling v_{jl}, and perform the trace $\mathrm{tr}_{l\neq j}$ over the degrees of freedom at sites $l \neq j$ using the action $S_{j\neq l}$. In the last step, products $c_{j_1}(t_1) c_{j_2}^{\dagger}(t_2)$ average to Green's functions $G_{j_1,j_2}^{[j]}$ of the lattice with a so-called cavity at site j, i.e., site j is excluded from the lattice. One can the re-exponentiate the terms (also to second order in v) in the form

$$Z = \mathrm{tr}_j \big[T_{\mathscr{C}}\, e^{iS_j^{eff}} \big], \tag{3.128}$$

which defines an effective action S_j^{eff} like (3.123), with

[11] For example, for the Hamiltonian (3.121), the hybridization function is the embedding self-energy discussed in Sect. 3.3.4: We have $\Delta(t,t') = \sum_p V_p(t) g_p(t,t') V_p(t')^*$, where $g_p(t,t')$ is the Green's function of an isolated bath orbital [(3.80)].

$$\Delta_j(t,t') = \sum_{j_1,j_2} v_{j,j_1}(t) G^{[j]}_{j_1,j_2}(t,t') v_{j_2,j}(t'). \tag{3.129}$$

Although this procedure was presented here for second oder, the result is already correct to all orders in the infinite-dimensional limit [51]. The hybridization has therefore the same form as the embedding self-energy of an open system [(3.80)], but with the bath Green's function taken to be the fully interacting Green's function of the lattice, and it intuitively represents processes of an electron hopping from site j to the rest of the lattice, propagating in the rest of the lattice, and hopping back at a different time.

3.5.2 Bethe Lattice

In addition to providing an intuitive interpretation of the DMFT action, the cavity method also gives a valuable simplification of the DMFT equations in a model situation which we will use in Sect. 3.6. Often one considers the DMFT equations on a so-called Bethe lattice (Fig. 3.5), which is an infinite tree in which each site has Z neigbors, but there are no closed loops. In the limit of $Z \to \infty$, we rescale the hopping $v_{jj'}$ between neighboring sites as $v_{jj'} = v_*/\sqrt{Z}$, analogous to (3.118). Furthermore, in this limit the Green's function $G^{[j]}_{ll}$ at a site l with site j removed becomes equal to the full Green's function G_{ll} (removing one site out of infinitely many does not make a difference), so that (3.129) for the hybridization function can be written in closed form

$$\Delta(t,t') = v_*(t) G(t,t') v^*_*(t'). \tag{3.130}$$

Here we have assumed for simplicity that all sites are equivalent. With this equation, the DMFT equations reduce to the determination of the Green's function $G(t,t')$ with a self-consistently determined hypridization function (3.130), without the need to explicitly solve a lattice Dyson equation. This equation is therefore frequently used to study properties of DMFT which are generic for any lattice. For the noninteracting case in equilibrium, the self-consistent solution corresponds to a quadratic equation

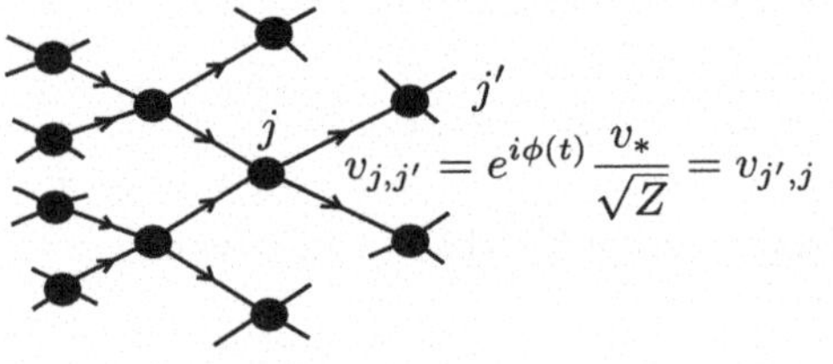

Fig. 3.5 A portion of the Bethe lattice with coordination number $Z = 4$. One can mimic the effect of an electric field by adding a Peierls phase $e^{i\phi(t)}$ to each bond, so that at each site there is an equal number of bonds with ingoing and outgoing field

$G(\omega + i0) = 1/[\omega + i0 - v_*^2 G(\omega + i0)]$, which gives the density of states (spectral function of the noninteracting model)

$$D(\epsilon) = \sqrt{4v_*^2 - \epsilon^2}/(2\pi v_*^2). \tag{3.131}$$

In non-equilibrium we are often interested in the effect of electric fields. In a tight-binding description like (3.117), time-dependent electromagnetic are most easily included by a Peierls substitution, which introduces the vector potential $\boldsymbol{A}(\boldsymbol{r}, t)$ as a phase factor in the hopping matrix elements,

$$v_{ij}(t) = v_{ij} \exp\left(-\frac{ie}{\hbar} \int_{\boldsymbol{R}_i}^{\boldsymbol{R}_j} d\boldsymbol{r}\, \boldsymbol{A}(\boldsymbol{r}, t) \right), \tag{3.132}$$

and adds a scalar potential term $e \sum_{i\sigma} \Phi(\boldsymbol{R}_i, t) c_{i\sigma}^\dagger c_{i\sigma}$ to the Hamiltonian (e is the charge of an electron). When we are interested in the action of optical or THz fields, the wavelength is much longer than the lattice spacing, and we can then assume that the field is space independent, so that $\boldsymbol{E}(t) = -\partial_t \boldsymbol{A}(t)$ and $\Phi = 0$. Including electric fields in multi-band models is more subtle, because the Peierls substitution does not describe any inter-band dipole couplings or Stark shifts of the Wannier orbitals. This fact becomes important for an ab-initio description of materials in the electric field. In these notes we focus on qualitative effects of the electric field which can be studied in a single-band model. The electric field relies on a geometric structure of the lattice, and the meaning of an electric field on a graph like the Bethe lattice is not a priori clear. Many DMFT simulations with electric fields have therefore been performed for a hypercubic lattice in the limit of infinite dimensions, where the electric field $\boldsymbol{E} = (E(t), E(t), ...) = -\partial_t (A(t), A(t), ...)$ points along the body-diagonal $(1, 1, ...)$ of the unit cell [58]. Each site j with coordination number Z has then $Z/2$ neigbors j' with hopping $v_{jj'} \propto e^{i\phi(t)}$ and $Z/2$ neigbors j' with hopping $v_{jj'} \propto e^{-i\phi(t)}$, where $\phi(t) = eaA(t)$ is the Peierls phase. The same situation of two types of nearest neighbor couplings at each site can also mimicked on the Bethe lattice (Fig. 3.5), leading to a self-consistency relation

$$\Delta(t, t') = \frac{v_*^2}{2} \left[e^{i\phi(t)} G(t, t') e^{-i\phi(t')} + e^{-i\phi(t)} G(t, t') e^{i\phi(t')} \right]. \tag{3.133}$$

Although clearly more realistic band structures can be (and have been) considered, we will use this simple and intuitive closed form self-consistency equation to obtain the results in Sect. 3.6.

3.5.3 Numerical Implementation and Impurity Solvers

In equilibrium, the self-consistent solution of the DMFT equations is usually achieved by an iterative procedure, where one starts from a guess for Σ, solves the lattice Dyson

(3.120) to obtain G_{ii}, inverts the impurity Dyson (3.125) to get Δ, and solves the impurity model with action (3.123) to get Σ. The DMFT equation are in essence a solution of the self-consistent lattice Dyson (3.120) with a highly nonlinear functional $\Sigma_{ii}[G]$ which is implicitly defined by solution of the impurity model. Because this functional is causal, i.e, $\Sigma(t, t')$ in some domain $t, t' \leq t_0$ can be determined from G in the same domain $t, t' \leq t_0$, the real-time DMFT equations can be solved step by step in time, as explained for the Dyson equation in Sect. 3.4.2. This part of a non-equilibrium DMFT implementation implies the same numerical cost as any solution of the Kadanoff-Baym equations, i.e., a polynomial scaling of the computer resources with the number of time-slices (see Sect. 3.4.2). The main conceptual and computational bottleneck in the DMFT framework is the solution of the impurity model, i.e., the determination of the Green's function or the self-energy for a given hybridization function. There is at present no multi-purpose approach which can solve the impurity problem in all parameter regimes, but a number of approximate methods which we summarize below. For a detailed description we refer to [29] and the original literature.

- **Weak-coupling expansions of the impurity self-energy** are numerically inexpensive (to lowest order, they imply the same scaling with the number of time-steps as the solution of the Dyson equation) and easy to implement [59]. In equilibrium, second-order perturbation theory (iterated perturbation theory, IPT) is known to extrapolate between weak and strong coupling when it is formulated in terms of the bare impurity Green's function (for the half-filled single-band model), and at least qualitatively reproduces the metal-insulator phase transition at intermediate coupling [51]. In non-equilibrium, however, the non-conserving nature of the bare expansion (see discussion at the end of Sect. 3.3.7) can become problematic [44], so that the use of IPT has so far remained restricted to the weak-coupling regime.
- **Quantum Monte Carlo techniques** stochastically sum the perturbation expansion in the hybridization function or the interaction to all orders. Quantum Monte Carlo is the standard approach in equilibrium at finite temperature, where it can give numerically exact results [60]. In real-time, however, the terms of the perturbation expansion become complex-valued, and Monte Carlo is limited to short times on the order of few hopping times due to the *dynamical sign problem* [61, 62]. There are interesting and fundamental problems to be studied in the short-time dynamics, such as dynamical phase transitions [63, 64], but for the study of the photo-induced dynamics few hopping times are usually not yet sufficient. Recent developments indicate ways to overcome the dynamical sign-problem [65], but have not been tested in the context of non-equilibrium DMFT.
- A **perturbation theory in the hybridization function** is currently the most flexible approach to study systems in the strongly interacting Mott regime (see Sect. 3.6). The approach can be formulated in conserving manner, and it can be extended to various settings, such as electron-phonon coupled systems [66], extended DMFT [67], bosonic DMFT [68], multi-orbital models [69], and cluster DMFT [70]. An implementation is described in [71]. However, only the first order (the so called non-crossing approximation, NCA) and to some extent the second

order are numerically feasible, restricting the approach to the Mott regime, and excluding the interesting range of correlated metallic phases at low-energy and close to half-filling.
- A promising novel direction is given by **Hamiltonian-based impurity solvers**. Here one tries to find a finite system which can accurately reproduce a given hybridization function $\Delta(t, t')$, such as the single-impurity Anderson model [c.f. (3.121) and footnote 11]. Representations have been derived for the steady state, using an impurity model in which the bath-sites are coupled to dissipative Lindblad terms [72] and for real-time [73]. In the latter case, wave-function propagation techniques based on matrix product states and the density matrix renormalization group (DMRG) can be used to compute the Green's function [74]. Such DMRG impurity solvers are only recently becoming feasible in the context of equilibrium DMFT [75, 76]. However, the numerical effort in these Hamiltonian-based techniques increases exponentially with the number of orbitals in the representation, which limits the frequency resolution in the steady state, and the accessible short times in the real-time formalism.

3.6 Photo-Doped Mott Insulators

3.6.1 The Mott Transition in DMFT

One of the hallmarks of strong electronic correlations in solids is the Mott transition. If the Coulomb energy between two valence electrons at the same atom exceeds the hopping energy to neighbouring sites on the lattice, electrons get localized and the system becomes insulating if there is an integer number of electrons per site. In contrast to a band-insulator, which has completely filled bands, electrons in the Mott-insulator still have active spin and orbital degrees of freedom, which is the origin of a variety of complex phases: The spin can order antiferromagnetically or ferromagnetically, and if there is more than one valence orbital, as in many transition metal compounds, orbital order can emerge, which is in turn strongly coupled to both magnetic properties and to the lattice [1]. Even in the disordered state, the behavior of the Mott insulator is strongly influenced by short-ranged magnetic and orbital correlations, and when electrons are doped into an Mott insulator, their coupling to these fluctuations may lead to yet new phases like superconductivity.

The understanding of the Mott transition requires an approach which can treat both local atomic correlations and the itinerant behaviour of electrons on the same footing. The development of DMFT has been instrumental in this respect [48]. Motivated by the experiments on photo-induced states in complex materials (Sect. 3.1), it has therefore been a natural step for the non-equilibrium extension of DMFT to look at the dynamics of Mott-insulators after various excitation protocols. In the last chapter of these notes, we will present an overview of the physics of photo-doped Mott insulators from the perspective of non-equilibrium DMFT. This will not be a review of the existing literature on the topic. Rather, we will present the main ideas and concepts in the simplest possible setting, which is the single-band Hubbard model (3.117) on a Bethe lattice (Sect. 3.5.2), where the DMFT impurity problem is solved within the leading order strong-coupling hybridization expansion (non-crossing approximation, Sect. 3.5.3). These single-band NCA simulations are relatively cheap and can after be done on a single computer. Although the NCA solution is not quantitatively accurate, in the parameter regime discussed here it can illustrate the main physical concepts, so that the results serve as a first reference for the reader entering the field. (NCA is known to underestimate the critical interaction for the Mott transition [71], but it gives qualitatively correct results at not too small U and finite temperature.) More accurate simulations, including more realistic band structure, high-order impurity solvers, or multi-orbital systems, are mentioned in the references.

To be precise, we adopt the setting of a Bethe lattice plus electric field, as described in Sect. 3.5.3, with self-consistency relation (3.133). The hopping $v_* = 1$ sets the energy scale, i.e., the noninteracting bandwidth is $W = 4$, and times are measured in units of $\hbar/v_*$. (For a typical energy scale $v_* = 1$eV, the unit of time is then 0.66fs.) Before going to non-equilibrium, let us first briefly recapitulate the physics of the equilibrium Mott transition in DMFT. Figure 3.6 shows the schematic phase diagram of the half-filled Hubbard model in DMFT [51]: If we restrict the calculation on the paramagnetic phase (i.e., translational invariance and equivalence between all sites is enforced in the simulation), on finds at low temperatures a phase transition between a gapless Fermi-liquid at small interactions, and a gapped Mott insulator

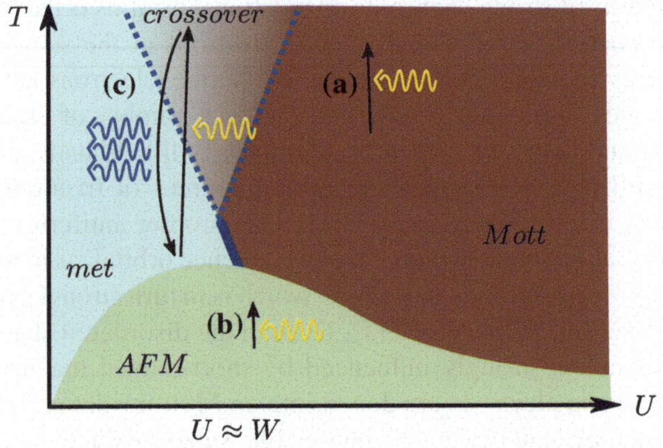

Fig. 3.6 Schematic phase diagram of the Hubbard model at half filling. On a bipartite lattice, the low-temperature phase is antiferromagnetically ordered. The bold blue line indicates the first order transition between a metal (light blue region) and the Mott insulator (brown region). The arrows show the various excitation processes discussed in the text: **a** Laser excitation of the isolated Mott phase and possible thermalization at high temperatures in the crossover regime (Sect. 3.6.2). **b** A similar excitation and thermalization in the Mott antiferromagnet (Sect. 3.6.3). **c** Laser excitation of the correlated metal close to the metal-insulator transition. In this case the system is coupled to an external phonon bath which can dissipate energy (blue wavy lines), and the relaxation back to the initial metallic phase is studied (Sect. 3.6.4)

at large U. The transition occurs roughly at $U \approx W$ (depending on the lattice), and becomes first-order at temperatures $T > 0$. The first order nature is evident, e.g., by a discontinuous decrease of the double occupancy from the metal to the Mott insulator. The phase transition line ends in a second-order end-point at temperature T^*, and for $T > T^*$ only a metal-insulator crossover is observed. Figure 3.7a shows the local (momentum-averaged) spectral function for various values of U at low temperatures. To be consistent with the time-dependent results below, these spectra have also been obtained within the NCA. In the correlated metallic phase ($U = 2.5, 3$), the spectrum features a peak at $\omega = 0$ which coexists with the Hubbard bands around $\omega = \pm U/2$. The peak corresponds to well-defined quasiparticles with a lifetime $1/\Gamma_k$ and a dispersion $\tilde{\epsilon}_k$ which give rise to a coherent peak $A_k(\omega) \sim Z_k \Gamma_k / [(\omega - \tilde{\epsilon}_k)^2 + \Gamma_k^2] + A_{inc}(\omega)$ in the spectrum. After summing over $\boldsymbol{k}$, the $\omega = 0$ peak in $A(\omega)$ indicates the bandwidth of the quasiparticle band, which becomes strongly renormalized with respect to the bare bandwidth close the the Mott transition. At the Mott transition, the quasiparticle band disappears, and a gap is opened ($U = 4, 5, 6$). It important to note that the quasiparticles are fragile against an increase of the temperature. Figure 3.7b shows the spectrum in the metallic phase ($U = 3$) for various temperatures. With increasing T, the quasiparticle peak is replaced by a dip in the spectral function. The high temperature state is a so-called bad-metal, in which coherent quasiparticles are absent and the scattering length becomes comparable to the lattice constant [77].

At low temperatures, the half-filled single-band Hubbard model shows a tendency towards antiferromagnetism. To study symmetry breaking in DMFT, we allow lattice sites to be inequivalent and solve one impurity model for each inequivalent site. This gives rise to a site-dependent (but still site-diagonal) self-energy. In the simplest case we consider a bipartite lattice (such as the Bethe lattice) with two sublattices

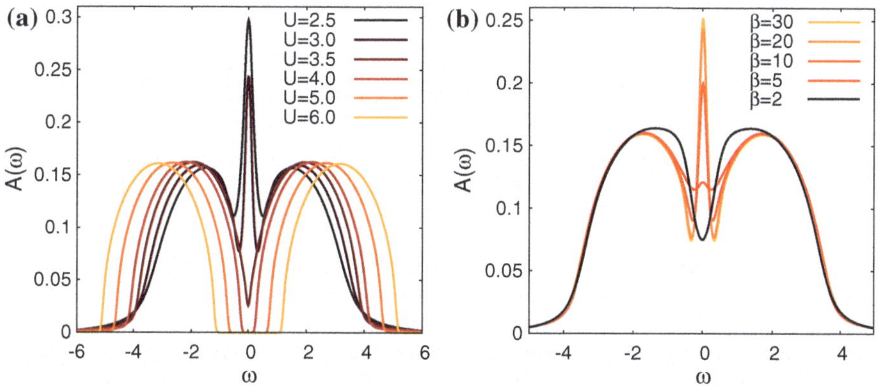

Fig. 3.7 Local spectral function $A(\omega)$ for the Hubbard model on a Bethe lattice, obtained within the non-crossing approximation. a) Spectra at low temperature ($\beta = 20$) for various values of U throughout the metal-insulator transition. b) Spectra at $U = 3$ for various temperatures, showing the crossover from a Fermi-liquid ($\beta = 30$) to a bad metal ($\beta = 2$)

A and B, and allow for a different self-energy on each lattice.[12] At strong-coupling, antiferromagnetism is a consequence of an super-exchange interaction $4v_*^2/U$ in the Hubbard model. If the Neel temperature T_N is above the critical end-point T_* of the Mott line, the first-order Mott transition is covered by the antiferromagnetic phase, but for band-structures which lead to magnetic frustration T_N can be suppressed far enough so that the Mott transition is uncovered. In phase diagrams of many real-materials, such as organic Mott insulators [78] or V_2O_3 [79] one in-fact observes a first order Mott transition.

3.6.2 Paramagnetic Phase - Dynamics of Photo-Excited Doublons

Relaxation of the Double Occupancy

The most straightforward excitation protocol is an electric field pulse with frequency $\Omega \approx U$, which will generate charge excitations in the paramagnetic Mott insulator, i.e. doubly occupied and empty sites. This question has been analyzed in [80] for the hypercubic lattice, using the strong-coupling expansion impurity solver to third order. Here we discuss the NCA results for the Bethe lattice, which give qualitatively similar results. Figure 3.8a shows the time-evolution of the doubly occupancy $d(t) = \langle n_\uparrow(t) n_\downarrow(t)\rangle$ during and after the excitation of the Hubbard model with a short field pulse. The field corresponds to a single-cycle laser pulse of duration $2\delta t$ with $\delta t = 2\pi/\Omega$, i.e., we take

$$A(t) = A_0 e^{-4(t-\delta t)^2/\delta t^2} \sin[(t-\delta t)\Omega)] \quad \text{for } 0 \leq t \leq 2\delta t \tag{3.134}$$

in (3.133). One can observe an increase of the double occupancy during the pulse, followed by a exponential approach of some final value.[13] While the pulse amplitude is chosen such that the increase of $d(t)$ during the pulse is comparable for the different values of U, the dynamics after the excitation is strongly dependent on U, which will be analyzed in the following.

---

[12] For the Bethe lattice, the self-consistency (3.130) is then modified to $\Delta_A(t,t') = v_*^2 G_B(t,t')$, because a site on the A sublattice has only neighbours on the B sublattice, and vice versa.

[13] There is a subtle difference between the double occupancy and the number of charge excitations ("doublons"). Due to virtual fluctuations $(\uparrow,\downarrow) \to (\uparrow\downarrow, 0) \to (\uparrow,\downarrow)$ of electrons on neighbouring sites, the doubly occupancy is never zero, not even at $T = 0$ when we can expect no charge excitations in the insulator. A rigorous distinction between a doublon number D and $d(t)$ is possible only in the limit $U \gg v_*$, by perturbatively projecting out terms in the Hamiltonian which change D, like in the t-J model. Spectroscopically, one can define D by the integrated weight in the upper band $\int_0^\infty d\omega\, N^<(\omega,t)$. As discussed in Sect. 3.2.3, this requires a separation of timescales, i.e., the number of charge excitations in the Hubbard model is only defined on timescales larger that $1/U$. In practice, the difference $d(t) - d(0)$ is a good measure for the excitation density.

As a first step, we will show that the dynamics observed after the pulse can indeed be interpreted as an approach of the thermal equilibrium state, as discussed in Sect. 3.2.1. Because the total energy of the system is conserved after the excitation,

$$\langle H(t)\rangle \equiv E_{tot} = \text{const.} \quad \text{for } t \geq 2\delta t, \tag{3.135}$$

thermalization would imply that the properties of the system eventually approach the properties of a system in equilibrium, at a temperature T_f which is defined such that the thermal expectation value $E_{th}(T_f)$ of the energy equals (3.135), $E_{tot} = E_{th}(T_f)$. We can therefore determine T_f by doing a series of equilibrium simulations at different temperatures, and comparing $E_{th}(T)$ with $\langle H(t)\rangle$. To analyze the thermalization in the Mott insulator, we then compute the thermal expectation value $d_{th}(T_f)$ for the different values of U in Fig. 3.8, and attempt a fit of the form

$$d(t) = d_{th}(T_f) + A\exp(-t/\tau), \tag{3.136}$$

see black dashed lines. (T_f is close to 0.5 for all curves.) One can see that the evolution of $d(t)$ is compatible with thermalization of the electronic system. The timescales, however, strongly depend on U, and range from few hopping times in the correlated metal ($U = 2.5$), to $\tau > 1000$ in the Mott phase ($U = 6$). Empirically, the dependence of the thermalization times on U can be described by the exponential dependence

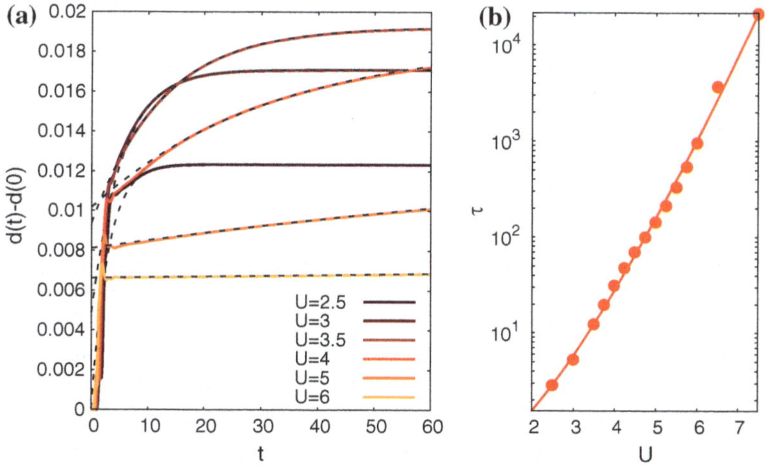

Fig. 3.8 **a** Time-evolution of the double occupancy in the Hubbard model on the Bethe lattice during and after a short pulse [(3.134)], for different values of U in the Mott insulator and crossover regime. The pulse duration $4\pi/U$ is less than about 6. The initial temperature is $T = 0.2$. Dashed lines are exponential fits, see discussion around (3.136). **b** Symbols: Relaxation times τ, obtained from the fits (3.136), as a function of U. The line shows the relation (3.137) with $\alpha = 0.302$

$$\tau(U) \sim \exp\left[\alpha \frac{U}{v_*} \log\left(\frac{U}{v_*}\right)\right], \tag{3.137}$$

with some numerical constant α (Fig. 3.8b). This result has an intuitive interpretation [81]: If a doublon and hole recombine, the energy U must be transferred to the remaining degrees of freedom. In the photo-excited Mott insulator, this implies mainly an increase of the kinetic energy of the remaining charge excitations (The energy of spin fluctuations is much lower). Because a single charge excitation can only take up an energy comparable to the bandwidth W in the system, the recombination of a doublon-hole pair for $U \gg W$ is a high-order process, for which the rate has been derived by a perturbative argument [82]: For $U \gg W$ one can transform the Hubbard model to a model in which doublons and holes numbers are conserved, with correction terms of order v_* in which doublons and holes either scatter without recombination, or recombine. The maximal change of the energy at each scattering processes of the first kinds is of the order of the bandwidth W. For simplicity, let us assume that $U = nW$, i.e., a transition from a state with $N + 1$ doublon-hole pairs to one with N doublon-hole pairs requires n such scattering processes and one recombination step, going through intermediate states with virtual energies U, $U - W$, $U - 2W$, etc. A Fermi-golden rule estimate of the transition rate then gives

$$v_* \frac{v_*}{U} \frac{v_*}{U - W} \cdots \frac{v_*}{W} = v_* \frac{1}{n!} \left(\frac{v_*}{W}\right)^n. \tag{3.138}$$

With the Stirling formula $n! \sim e^{n \log(n)}$ and $n = U/W$, we arrive at (3.137). A similar exponentially long lifetime is generally expected when a high-energy particle decays into a shower of low-energy excitations. In the context of the Hubbard model, the decay rate (3.137) was first measured in a system of ultra-cold atoms, where doublons are generated by a modulation of the hopping amplitude [81]. In condensed mater, the life-times of photo-excited carriers in Mott insulators can range to picoseconds [83–85], which is thousands of hopping times and thus compatible with the described bottleneck. However, in real materials the decay mechanisms can involve other degrees of freedom, such as phonons or spins. Furthermore, at low excitation densities there are simply no other charge carriers present which could release kinetic energy. In this case the decay into spin-fluctuations has been discussed as a possible mechanism for the doublon decay [86]. In organic materials, on the other hand, there are also phonons at relatively high energies ω_0 (molecular vibrations), so that a multi-phonon decay can become important (the timescale is then exponential in U/ω_0 rather than U/W [85, 87]).

Relaxation of the Spectral Function

We now focus on the regime of small U, where rapid thermalization is observed in the double occupancy. As discussed in Sect. 3.2.1, thermalization would imply also that eventually correlation functions satisfy the fluctuation dissipation relation.

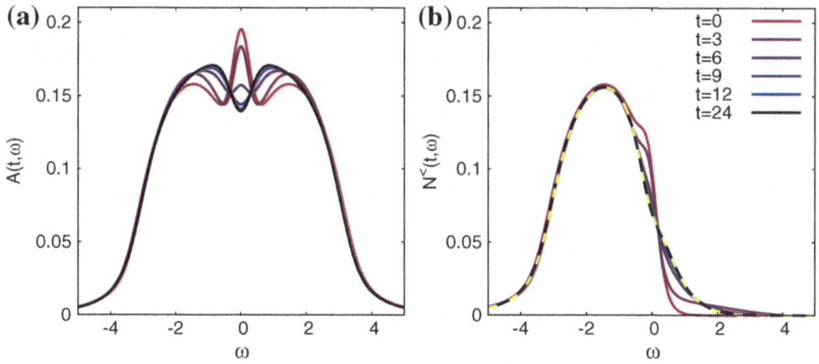

Fig. 3.9 a) Spectral function [(3.139)] for $U = 2.5$, $\beta = 5$ and the same excitation protocol as in Fig. 3.8, plotted for different times t (legend in right panel). b) Occupation function (3.140) for the same parameters. The yellow dashed line (which lies on top of the curve $t = 24$) shows $A(t, \omega) f(\omega, T_f)$, with $T_f = 1/1.967$ determined from the total energy

Figure 3.9 shows the spectral function and occupation function, which we define by the backward Fourier transform (with some sufficiently large cutoff s_{max})

$$A(\omega, t) = -\frac{1}{\pi}\text{Im}\int_0^{s_{max}} ds\, G^R(t, t-s)e^{i\omega s} \tag{3.139}$$

$$N^<(\omega, t) = -\frac{1}{\pi}\text{Im}\int_0^{s_{max}} ds\, G^<(t, t-s)e^{i\omega s} \tag{3.140}$$

for various times after the excitation. The data have been obtained for $U = 2.5$, and the same excitation as in Fig. 3.8. Both spectral function and occupation function become independent of t within a time of the order of the relaxation time τ of the double occupancy [c.f. Fig. 3.8b]. At the latest time ($t = 24$ in Fig. 3.9), we can indeed verify that the fluctuation theorem is satisfied in the form $N^<(\omega) = A(\omega) f(\omega, T_f)$. The best fit temperature T_f coincides with the value $T_f = 1/1.967$ obtained from the total energy. In addition to the observation of thermalization, it is interesting to note that not only the occupation of states, but also the spectral function completely changes during the evolution: In the initial state ($\beta = 5$) there is still a reminiscence of the quasiparticle peak, while in the final state the system is in a bad-metallic state, where the spectrum has a pseudo-gap instead of the quasiparticle peak. This shows that the Fermi liquid can de destroyed in only few electronic hopping times. In contrast, reforming the quasi-particles is a process which takes much longer, as discussed Sect. 3.6.4.

Impact Ionization

We have formulated the bottleneck for thermalization of doublons in terms of doublon-hole recombination, but one can see in Fig. 3.8 that the double occupancy actually increases after the pulse instead of decreasing. This simply means that the kinetic energy of the photo-excited state is initially too high, and thermalization is achieved by taking kinetic energy from one or more charge excitations to generate an additional doublon-hole pair. The generation of additional charge-carriers in the relaxation process is similar to an Auger process in atomic physics, and can be called *impact ionization* [88] in analogy to the related process in semiconductors. When the bandwidth of the Hubbard bands is larger than the gap [as in the vicinity of the Mott transition, e.g., $U = 4$ in Fig. 3.7a], impact ionization can happen when a single doublon decays within the upper band.

The process is visible also in the evolution of the occupation function. Figure 3.10 shows the occupied density of states (3.140) in the upper Hubbard band for $U = 4$ and the same excitation protocol as in Fig. 3.8. One can see an increase of the weight at small energies, and a decrease at high energies, which indicates loss of kinetic energy of some photo-excited doublons. The increase at low energy is larger than the decrease, which shows that kinetic energy is used to excite additional carriers. The impact ionization is thus clearly distinguishable from intra-band relaxation via spins and phonons, where the total weight remains constant while it is redistribute from higher to lower energies. In the ideal case, when the relaxation of each doublon from the high-energy window generates precisely one doublon-hole pair, one would expect the increase of the weight at low energies to be larger by a factor 3 than the decrease at high energies. This argument works surprisingly well, even though the energy of a doublon is not really well defined (the width of the peaks in the momentum-resolved spectral $A_k(\omega)$ function is of order one): In the present case, the increase of weight from $t = 18$ to $t = 60$ in the interval $0.5 \leq \omega \leq 2$ is larger by approximately a factor 3.1 than the decrease in the interval $2 \leq \omega \leq 4$, hinting even at double impact ionization processes.

Impact ionization processes are interesting in the context of photo-voltaic devices [89]: Usually, a photon creates a single charge excitation, which then quickly relaxes to the bottom of the band via scattering with phonons. The usable energy per photon is

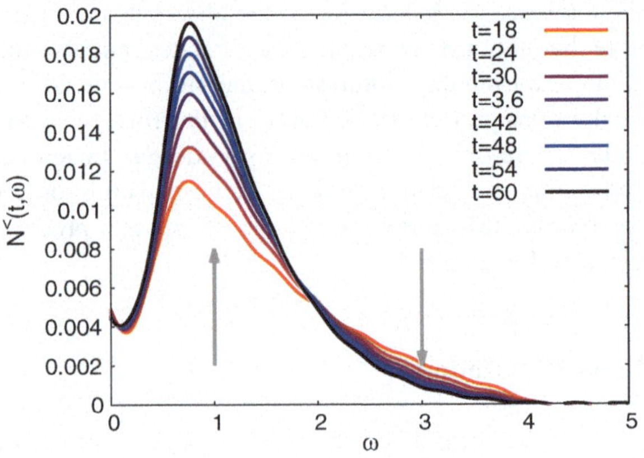

Fig. 3.10 Impact ionization: Occupation function (3.140) for $U = 4$ and the same excitation protocol as in Fig. 3.8, plotted for different times t. The arrows indicate increase and decrease of weight at low and high energies respectively

therefore not given by $\hbar\omega$, but only by the gap Δ. An average of all possible processes would give an efficiency of about 33% for a photo-voltaic device operating on the spectrum of sunlight, when the gap is chosen optimally (Shockley-Queisser limit). In contrast, if the relaxation involves impact ionization, the final state can have *two or more* electrons at the energy Δ, which can increase the efficiency beyond the Schockley-Queisser limit. Impact ionization may become relevant if the rate is faster than competing intra-band relaxation processes, such as phonons. The DMFT results show that impact ionization processes in small-gap Mott insulators indeed can occur on the time-scale of a few hoppings [88], which is much faster than typical electron phonon relaxation times. On the other hand, in Mott insulators short-range spin fluctuations provide a particularly fast intra-band relaxation mechanism (Sect. 3.6.3), and it remains to be seen to what extent impact ionization plays an important role in real Mott materials.

3.6.3 Antiferromagnetic Case

At half-filling and on a bipartite lattice, the DMFT solution of the Hubbard model shows an extended antiferromagnetic phase at low temperature. The non-equilibrium dynamics of this phase has been investigated both for weak-coupling [18], where it is well described as a Slater mean-field antiferromagnet, and for strong-coupling, where the Heisenberg model is the limiting description [90]. Similar to the analysis of photo-excitation in the paramagnetic phase, we now investigate the behavior of the antiferromagnet after excitation with an electric field pulse, with a focus on the Mott insulator at large U. Just like in the paramagnetic phase, the application of an electric field pulse (3.134) leads to an impulsive increase of the double occupancy, which for large U does not thermalize on the timescales accessible in the simulation. We can take the increase $n_{ex} = d(2\delta t) - d(t)$ of $d(t)$ during the pulse as a measure for the excitation density, and analyze the effect of this non-thermal doublon and hole population on the antiferromagnetic order. The results for the photo-induced state at interaction U are qualitatively similar to the behavior after a quench of the interaction from some value U_i to U [90]. In both cases, the long-time behavior of the staggered order parameter is governed by the density of non-thermalized doulons. In the quench, the latter is determined by the quench amplitude $\Delta U = |U - U_i|$.

Figure 3.11a shows the time-evolution of the antiferromagnetic order parameter

$$m(t) = \langle c^\dagger_{j\uparrow} c_{j\uparrow} - c^\dagger_{j\downarrow} c_{j\downarrow} \rangle_j \text{ on } A - \text{sublattice} \tag{3.141}$$

in the Hubbard model on the Bethe lattice after excitation with an electric field pulse. The order parameter shows a rapid decrease after the pulse. For excitation densities $n_{ex} \lesssim 0.015$, it saturates to a finite value, while for $n_{ex} \gtrsim 0.015$, we observe complete melting of long-range order. In both cases, the approach of the final state in the simulated time-window can be described with an exponential function $m(t) = m_\infty + Ae^{-\gamma t}$. For weak excitation densities, the fit becomes less reliable because there are damped amplitude-oscillations on $m(t)$. The plot of the final value m_∞ as a function of the excitation density shows a relatively sharp threshold for the melting of the antiferromagnetic phase (Fig. 3.11b). At the threshold for melting, there is slowdown of the melting time, consistent with a critical slowdown at a second-order phase transition. For the simulation of the dynamics after a quench of size ΔU [90], where somewhat longer times have been studied, a critical slowdown $\gamma \sim |\Delta U - \Delta U_c|$ was observed, where ΔU_c is the critical quench amplitude for melting the antiferromagnetic phase. Analogous to the discussion for the paramagnetic phase, we can evaluate the temperature T_f which would be reached after thermalization. The corresponding value of the order parameter in the thermalized state, $m_{th}(n_{ex})$ is found to be always lower than the value of m in the photo-excited state (Fig. 3.11b). If the system would thermalize, long-range order would melt already for $n_{ex} \lesssim 0.01$, while it remains for $n_{ex} \lesssim 0.015$ in the non-thermalized state. The intermediate state is expected to relax only on the exponentially long timescale on which doublons and holes recombine by a transfer of the energy to the magnetic sector.

The melting of the antiferromagnetic order can be understood in simple terms [93, 94]: A hole which is moving in the antiferromagnetic spin background induces a spin-flip in every hopping process. Because each spin flip increases the energy in the magnetic sector by an amount $\mathcal{O}(J_{ex}) \sim v_*^2/U$, this implies an ultra-fast transfer of kinetic energy from photo-doped electrons to the spins. The latter becomes also visible in the spectrum: Figure 3.12b shows the evolution of $N^<(\omega, t)$ in the upper Hubbard band, for a intermediate excitation density. After the pulse, the population

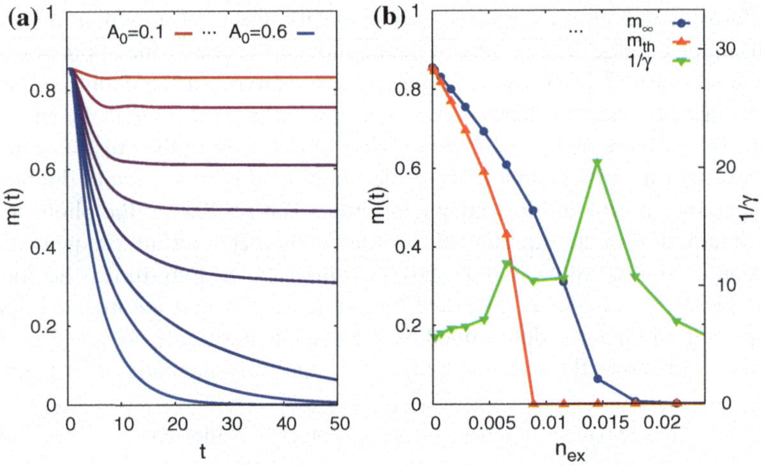

Fig. 3.11 a) Time-evolution of the antiferromagnetic order parameter $m(t)$ in the Hubbard model on the Bethe lattice ($U = 5$, $\beta = 20$), after excitation with an electric field pulse (3.134). The different curves correspond to different amplitudes, with $A_0 = 0.05$ to 0.6 from top to bottom. The corresponding excitation densities range up to $n_{ex} \approx 0.025$. b) Final value m_∞ and the relaxation time $1/\gamma$, obtained by fitting the curves in panel a) with $m(t) = m_\infty + Ae^{-\gamma t}$ in the time window $t \geq 15$, plotted against the excitation density. m_{th} would be the order parameter after thermalization

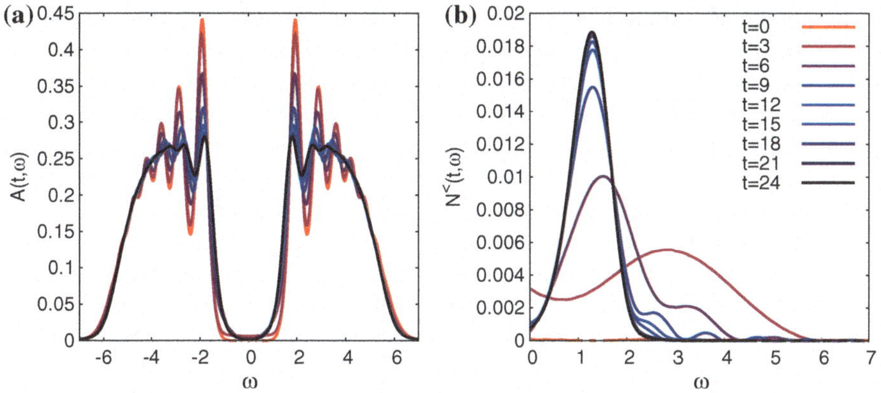

Fig. 3.12 **a** Spectral function [(3.139)] for the parameters of Fig. 3.11, with $A_0 = 0.3$ ($n_{ex} \approx 0.007$), plotted for different times (see legend in right panel; $s_{max} = 20$ in (3.139)). **b** Occupation function [(3.140)] for the same parameters, plotted in the upper Hubbard band only

in the upper band redistributes to lower energies. In contrast to the impact ionization, the population in the upper band is conserved (the relative change of the integrated weight $\int_0^8 d\omega N^<(\omega, t)$ for times after the pulse is less than one per cent). This shows that the scattering of electrons with spins provides a ultra-fast mechanism for "intra-band relaxation". We note that again not only the occupation function, but also the spectrum changes after the pulse (Fig. 3.12a). In particular, we observe a slight filling of the gap, similar to the paramagnetic case, and a reduction of the side-bands on the Hubbard bands, which reflect the motion of a doublon or hole in the presence of antiferromagnetic order [91].

The results on the dynamics in Mott antiferromagnets open a number of question and topics, which can only briefly be summarized here:

- **Implications of intra-band scattering**: Due to the strong coupling of electrons and spins, the latter act as a very efficient heat bath for the electrons. This strongly influences transport properties, such as the motion of photo-excited carriers close to surfaces and interfaces where strong-internal fields can be found [92]. The mechanism is also in competition to the impact ionization discussed in the previous section. It should be noted that the rapid energy transfer to spins is also possible in the paramagnetic phase, where spins have no long-range order, but are short-range and short-lived. Such processes have been studied with exact diagonalization techniques [93, 94]. For a $t - J$ model with parameters comparable to cuprates, the relaxation times $\tau \sim 10 - 20 fs$ are compatible with the timescales for the initial saturation of the optical response in cuprates after a few-fs laser excitation [95]. Short-range fluctuations cannot be captured in conventional single-site DMFT, but they have been studied using cluster extensions (the dynamical cluster approximation) [70], where again ultra-short relaxation timescales, comparable to the previous exact diagonalization results, have been found.

- **The nature of the photo-doped state**: The second question regards the characterization of the so-called photo-doped state, i.e., the non-thermal state which is obtained after the kinetic energy has been transferred to the spin sector. One can show that although this photo-doped state is not thermal, its properties are determined by only one additional control parameter, which is the excitation density [90]: Firstly, one can show that the spectral function of the photo-excited state with excitation density n_{ex} and order parameter m_∞ is well reproduced by the spectral function of a *doped system* in which the density of carriers $n-1$ (which are either doublons or holes) equals $2n_{ex}$, i.e., the density of doublons *and* holes in the photo-excited state [90]. In contrast, it differs from the spectral function of the half-filled system at any temperature. Furthermore, it was shown that also the magnetic exchange interaction (which governs spin precessional motion) is modified in the photo-excited state to the value obtained in an equilibrium state with the same density of carriers [96]. The photo-doped state in the single-band antiferromagnet can therefore be viewed as the simplest possibility in which the photo-excited state of a correlated system is changed (after some primary electronic relaxation) to a different state which is basically controlled by a single additional control parameter, similar to what has been empirically suggested in an experiment on a manganite [97]. In systems with more than one orbital, there is a large playground to manipulate the occupations of different types of doubly occupied and multiply occupied states [69] and use this to modify the effective interactions which govern spin and orbital order.

3.6.4 The Buildup of the Fermi Liquid

In Sect. 3.6.2 we have seen that quasiparticles can be destroyed within a few hopping times, when a photo-excited system close to the Mott transition thermalizes in the high-temperature bad metallic state. It is therefore an obvious question to ask on which timescale the inverse process can happen, and a Fermi liquid can be formed out of the bad metal by cooling the system. As the existence of well-defined quasiparticles is one of the most important concept in many-body physics, this question can be considered of rather fundamental interest. The problem has been addressed in [98] and [99], and in the following section we will mainly discuss the results of [99]. The cooling of the photo-excited state requires coupling to a heat bath, which is in principle always present in a real solid, e.g., by coupling electrons to phonons. If electrons would thermalize within few hopping times at all temperatures ranging from the bad metal to the onset of the Fermi liquid, the timescale for forming the Fermi liquid would be limited only by the time in which the energy can be transferred to the bath (assuming the latter is considerably longer than the hopping time). However, we will observe the opposite, i.e., there is a much slower bottleneck of electronic origin which controls the reformation of the Fermi liquid.

Figure 3.13a shows the evolution of the spectral function (3.139) in the Hubbard model on the Bethe lattice after a short pulse. In contrast to the results in the previous

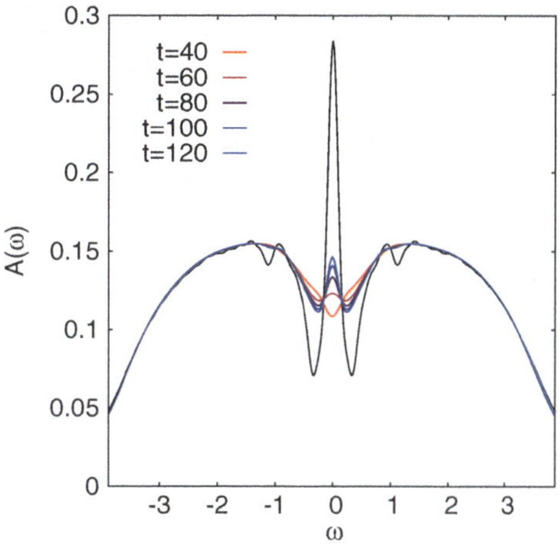

Fig. 3.13 a) Spectral function [(3.139)] of the Bethe lattice at $U = 3.1$, coupled to a phonon heat bath (see text). The black solid line is the spectrum in equilibrium at the temperature of the bath ($\beta = 30$)

sections, we now study a model which is in contact to a heat bath. To describe dissipation, the lattice Hamiltonian is coupled to additional degrees of freedom, which are traced out and simply give a self-energy contribution Σ_{bath}. This procedure can be incorporated within DMFT as long as the bath is coupled locally to each site. Two main forms of dissipation have been considered [29], in which the bath is either represented by noninteracting Fermions [then the dissipative self-energy has the form (3.80)], or by a continuum of bosonic degrees of freedom. In general, the precise mechanism of dissipation should be irrelevant for the physics, otherwise the bath degrees must be explicitly included in the time-evolution. In order for the bosons to act as a heat bath, we have to neglect the back-action of the electrons on the phonons, i.e., we take Σ_{bath} to be the simple first-order diagram of a local electron phonon interaction,

$$\Sigma_{\text{bath}} = \lambda^2 G(t,t') D_0(t,t'), \tag{3.142}$$

$$D_0(t,t') = -i \int d\omega\, \Gamma(\omega)[\theta_{\mathscr{C}}(t,t') + b(\omega)]\, e^{i\omega(t'-t)}, \tag{3.143}$$

where G is the fully interacting local electron Green's function, D_0 is the propagator for free bosons with a density of states $\Gamma(\omega)$, and λ measures the coupling strength. The occupation function of the bosons is kept in equilibrium (no back action), i.e., it is given by $b(\omega) = 1/(e^{\beta\omega} - 1)$ with inverse temperature β. This first-order approximation is justified because the interpretation of the bosons in terms of a heat-bath is only meaningful for weak coupling. Here we consider a bosonic bath with frequencies ranging to $\omega = 1$ and $\lambda = 0.5$. The coupling is strong enough to allow for a rapid energy dissipation, but still so weak that only relatively minor differences occur in the equilibrium spectrum due to the coupling to the environment (compare the equilibrium spectra in Figs. 3.13a and 3.7a).

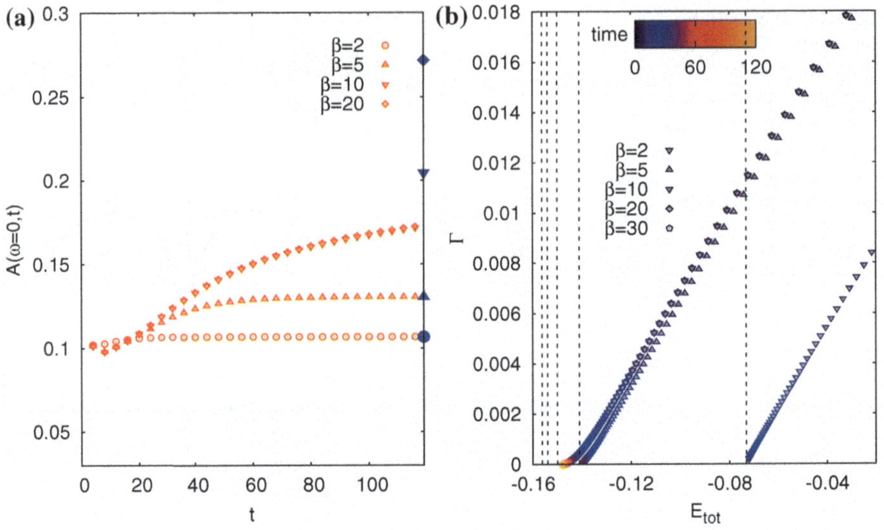

Fig. 3.14 a) The evolution of the height of the quasiparticle peak $A(\omega = 0, t)$ for different bath temperatures. The blue filled symbols on the right indicate the equilibrium value of $A(\omega)$ at the respective bath temperatures. b) The rate for energy transfer, $\Gamma(t) = |dE_{tot}/dt|$, plotted against the time-dependent energy $E_{tot}(t)$ for different bath temperatures. Note that the data points for $\beta \geq 10$ basically lie on top of each other. The time increases from higher to lower energies (see color map). Vertical dashed lines indicate the equilibrium energies for inverse temperatures $\beta = 30, 20, 10, 5, 2$ (from left to right)

In Fig. 3.13a one can see that after the excitation the quasiparticle peak does not reform within more than hundred hopping times, although electronic thermalization at large temperature happens in few hopping times.[14] To see that the origin for the slow evolution is an electronic bottleneck and not restricted by the coupling to the bath, we consider the evolution of the spectral function $A(\omega, t)$ at $\omega = 0$, i.e., the height of the quasiparticle peak, for different values of the bath temperature (Fig. 3.14a). The equilibrium value of $A(\omega = 0, t)$ is indicated by the filled symbols on the right vertical axis. For high bath temperatures, the initial state is rapidly recovered, but for $\beta \gtrsim 10$, the evolution becomes basically independent of the bath temperature. A more clear evidence of an electronic bottleneck is given by the rate of energy transfer to the bath, which is defined as $\Gamma(t) = |dE_{tot}/dt|$, where $E_{tot}(t) = \langle H(t) \rangle$ is the total energy of the system alone. An exponential relaxation of E_{tot} to some final value $E_{tot}(\infty)$ would give $\Gamma(t) \propto |E_{tot}(t) - E_{tot}(\infty)|$. Figure 3.14b shows that this relation holds quite well for high bath temperatures ($\beta \lesssim 5$), and $E_{tot}(\infty)$ is given by the energy $E_{tot}(\beta)$ of the system at the bath temperature. For lower bath temperatures, however, the energy transfer rate to the bath freezes out below some energy E^*_{tot}

[14]The excitation is a short modulation of the hopping amplitude instead of the phase [99]. However, as the excited state with energies corresponding $\beta \approx 1$ thermalizes within few hopping times, the precise excitation mechanism does not matter for the long-time behavior.

which is larger than $E_{tot}(\beta)$. For early times the dynamics seems to extrapolate to E^*_{tot} instead of the final energy. If we take E^*_{tot} to be the linear extrapolation of Γ in the range $-0.12 \leq E_{tot} \leq -0.04$ to zero, we have $E_{tot} \approx -0.143$, which corresponds to $T_* \approx 1/6$. The latter lies in the temperature range where the onset of a peak in $A(\omega)$ at $\omega = 0$ is observed in equilibrium.

A possible explanation for the slowdown of the electronic relaxation would be that in the Fermi liquid a new timescale emerges, i.e., the lifetime of the quasiparticles, which increases like T^{-2} for decreasing temperature. However, for $T \approx T_*$, quasiparticles are not yet well formed. The spectral function $A_{\epsilon_F}(\omega)$ of an electron at the Fermi energy [for the particle-hole symmetric Bethe lattice, we have $\epsilon_F = 0$, and $A_{\epsilon_F}(\omega) = -\frac{1}{\pi}\text{Im}[\omega + i0 - \Sigma^R(\omega)]^{-1}]$ has only a relatively broad peak for $T = T_*$, with a width of order one. This clearly shows that there is a different bottleneck involved in the formation of the Fermi liquid.

A possible explanation of the bottleneck was identified in the work by Sayyad et al. [99], based on a slave-rotor representation of the Hubbard model [100]. In the slave-rotor representation, the Fermions are represented as "composites" of a quantum rotor $\theta \in [0, 2\pi)$ and a fermion f (the spinon). The impurity operators (c_σ, $c_\sigma^\dagger$) are substituted by $c_\sigma^\dagger = f_\sigma^\dagger e^{i\theta}$. A constraint $L = \sum_\sigma f_\sigma^\dagger f_\sigma - 1$ on the angular momentum $L = i\partial_\theta$ of the rotor removes unphysical states from the Hilbert space. With this, the rotor angular momentum of the rotor represents the electron charge, $L^2 = (c_\uparrow^\dagger c_\uparrow - \frac{1}{2})(c_\downarrow^\dagger c_\downarrow - \frac{1}{2})$. Even close to the metal-insulator transition, the DMFT solution can be well approximated can by a perturbative solution of the coupled spinor and rotor degrees of freedom. In [99] it was observed that the spinon lifetime τ_{spinon} becomes large around T^*, and can describe the bottleneck (the time-dependence of $A(\omega = 0, t)$ for different values of the interaction roughly follow a universal curve when plotted against t/τ_{spinon}). In this sense the non-equilibrium dynamics reveales the composite nature of the electron which remains hidden in equilibrium. This viewpoint may be seen as a consequence of a mathematical trick of separating the electron into spinor and rotor, but it clearly shows that a description of the dynamics in terms of electronic quasiparticles is not a good starting point. Ultra-short quasi-particle lifetimes, i.e, rapid quasiparticle scattering, do not at all imply rapid thermalization. This results can be expected to carry over to more complex systems than the single-band model.

The timescale for forming the Fermi liquid is also of practical importance. In a Mott insulator, one can imagine that by coupling to a bath, the doublons (which have an exponentially long long lifetime, see Sect. 3.6.2) relax to the bottom of the band and form a Fermi liquid before recombination. However, analogous to the behavior described above, the numerical data show that although a bath can efficiently reduce the kinetic energy of the doublons in a photo-excited Mott insulator, the lifetime of the electronic excitations remains similar to that of a bad metal [98], indicating a bottleneck for the formation of a Fermi liquid. This determines the optical properties of the metallic state in a photo-doped Mott insulator.

3.7 Concluding Remarks

In these notes we have described the technical foundations for a theoretical description of correlated electron systems out of equilibrium using non-equilibrium DMFT. Clearly, this chapter is not a review of non-equilibrium DMFT, and does not attempt to cover all the possible directions of exploring non-equilibrium states. For example, a large field of research which has been left out includes the so-called Floquet engineering of Hamiltonians by periodic driving [101], where DMFT provides a framework to study directly the driven dissipative state [102]. Also, we note that most studies so far rely on simple single-band models with ad-hoc parameters. In contrast, many correlated materials have more than one active orbital, so that their physics is governed by the interplay of orbital order, spin-order, and the strong coupling to the lattice. Consequences of the multi-orbital nature have already been addressed within the simpler Gutzwiller approximation [103, 104], and within first multi-orbital DMFT [69, 105]. Furthermore, the feedback of non-equilibrium states on the interaction parameters provides another route to explore interesting non-equilibrium phenomena [106]. The future goal will be to understand how one can dynamically address the various degrees of freedom to steer a complex system along a nontrivial reaction path into novel transient or hidden phases, or dynamically stabilize non-equilibrium states.

Acknowledgements These notes are based on a series of lectures given at the *XX. Training Course in the Physics of Strongly Correlated Systems*. The author would like to thank Ferdinando Mancini and Roberta Citro for organizing this fruitful summer school over the years, and the participants of the training course for actively contributing to a productive and enjoyable atmosphere.

References

1. M. Imada, A. Fujimori, Y. Tokura, Metal Insulator Transitions. Rev. Mod. Phys. **70**, 1039 (1998)
2. E. Dagotto, Complexity in strongly correlated electronic systems. Science **309**, 257 (2005)
3. C. Giannetti, M. Capone, D. Fausti, M. Fabrizio, F. Parmigiani, D. Mihailovic, *Ultrafast optical spectroscopy of strongly correlated materials and high-temperature superconductors: a non-equilibrium approach* Adv. in Phys. **65**, 58 (2016)
4. S. Dal Conte et al., Disentangling the electronic and phononic glue in a high-Tc superconductor. Science **335**, 1600 (2012)
5. T. Rohwer et al., Collapse of long-range charge order tracked by time-resolved photoemission at high momenta. Nature **471**, 490 (2011)
6. M. Rini, R. Tobey, N. Dean, J. Itatani, Y. Tomioka, Y. Tokura, R.W. Schoenlein, A. Cavalleri, Control of the electronic phase of a manganite by mode-selective vibrational excitation. Nature **449**, 72 (2007)
7. D. Fausti, R.I. Tobey, N. Dean, S. Kaiser, A. Dienst, M.C. Hoffmann, S. Pyon, T. Takayama, H. Takagi, A. Cavalleri, Light-Induced Superconductivity in a Stripe-Ordered Cuprate. Science **331**, 189 (2011)
8. W. Hu, S. Kaiser, D. Nicoletti, C.R. Hunt, I. Gierz, M.C. Hoffmann, M. Le Tacon, T. Loew, B. Keimer, A. Cavalleri, Nature Materials **13**, 705 (2014)

9. Mitrano et al., Possible light-induced superconductivity in K3C60 at high temperature. Nature **530**, 461 (2016)
10. H. Ichikawa et al., Transient photoinduced 'hidden' phase in a manganite Nat. Materials **10**, 101 (2011)
11. L. Stojchevska, I. Vaskivskyi, T. Mertelj, P. Kusar, D. Svetin, S. Brazovskii, D. Mihailovic, Ultrafast switching to a stable hidden quantum state in an electronic crystal. Science **344**, 177 (2014)
12. S. Mor et al., Ultrafast electronic band gap control in an excitonic insulator Phys. Rev. Lett. **119**, 086401 (2017)
13. A. Polkovnikov, K. Sengupta, A. Silva, M. Vengalattore, Colloquium: nonequilibrium dynamics of closed interacting quantum systems. Rev. Mod. Phys. **83**, 863 (2011)
14. J. Berges, S. Borsańyi, C. Wetterich, Prethermalization. Phys. Rev. Lett. **93**, 142002 (2004)
15. M. Moeckel, S. Kehrein, Interaction quench in the hubbard model. Phys. Rev. Lett. **100**, 175702 (2008)
16. M. Kollar, F.A. Wolf, M. Eckstein, Generalized Gibbs ensemble prediction of prethermalization plateaus and their relation to nonthermal steady states in integrable systems. Phys. Rev. B **84**, 054304 (2011)
17. B. Sciolla, G. Biroli, Quantum quenches, dynamical transitions, and off-equilibrium quantum criticality. Phys. Rev. B **88**, 201110(R) (2013)
18. N. Tsuji, M. Eckstein, Ph Werner, Nonthermal antiferromagnetic order and nonequilibrium criticality in the Hubbard model. Phys. Rev. Lett. **110**, 136404 (2013)
19. D. Golež, Ph Werner, M. Eckstein, Photo-induced gap closure in an excitonic insulator. Phys. Rev. B **94**, 035121 (2016)
20. G.D. Mahan, *Many-Particle Physics* (Kluver Academic, New York, 2000)
21. L.V. Keldysh, Diagram technique for nonequilibrium processes. JETP **20**, 1018 (1965)
22. L.P. Kadanoff, G. Baym, *Quantum Statistical Mechanics* (Perseus Books, Cambridge, 1962)
23. L. Bányai, Q.T. Vu, B. Mieck, H. Haug, Ultrafast quantum kinetics of time-dependent RPA-screened coulomb scattering. Phys. Rev. Lett. **81**, 882 (1998)
24. M. Bonitz (ed.), *Progress in Nonequilibrium Green's Functions* (World Scientific, Singapore, 2000)
25. K. Balzer, M. Bonitz, *Nonequilibrium Green's Functions Approach to Inhomogeneous Systems* (Spinger, Heidelberg, 2013)
26. M. Babadi, E. Demler, M. Knap, Far-from-equilibrium field theory of many-body quantum spin systems: Prethermalization and relaxation of spin spiral states in three dimensions. Phys. Rev. X **5**, 041005 (2015)
27. M. Sentef, A.F. Kemper, B. Moritz, J.K. Freericks, Z.-X. Shen, T.P. Devereaux, Examining electron-boson coupling using time-resolved spectroscopy. Phys. Rev. X **3**, 041033 (2013)
28. Y. Murakami, Ph Werner, N. Tsuji, H. Aoki, Interaction quench in the holstein model: thermalization crossover from electron- to phonon-dominated relaxation. Phys. Rev. B **91**, 045128 (2015)
29. H. Aoki, N. Tsuji, M. Eckstein, M. Kollar, T. Oka, Ph Werner, Nonequilibrium dynamical mean-field theory and its applications. Rev. Mod. Phys. **86**, 779 (2014)
30. M. Srednicki, Chaos and quantum thermalization. Phys. Rev. E **50**, 888 (1994)
31. J.M. Deutsch, Quantum statistical mechanics in a closed system. Phys. Rev. A **43**, 2046 (1991)
32. M. Rigol, V. Dunjko, M. Olshanii, Thermalization and its mechanism for generic isolated quantum systems. Nature **452**, 854 (2008)
33. R. Kubo, M. Toda, N. Hashitsume, *Statistical Physics II: Nonequilibrium Statistical Mechanics* (Springer, Heidelberg, 1991)
34. F. Randi, M. Esposito, F. Giusti, F. Parmigiani, O. Misochko, D. Fausti, M. Eckstein, *Probing the fluctuations of the optical properties in time-resolved spectroscopy*, arXiv:1705.08523 (to appear in Phys. Rev. Lett.)

35. J.K. Freericks, H.R. Krishnamurthy, Th Pruschke, Theoretical description of time-resolved photoemission spectroscopy: application to pump-probe experiments. Phys. Rev. Lett. **102**, 136401 (2009)
36. M. Eckstein, M. Kollar, Measuring correlated electron dynamics with time-resolved photoemission spectroscopy. Phys. Rev. B **78**, 245113 (2008)
37. F. Randi, D. Fausti, M. Eckstein, Bypassing the energy-time uncertainty in time-resolved photoemission. Phys. Rev. B **95**, 115132 (2017)
38. A. Kamenev, *Field Theory of Non-Equilibrium Systems* (Cambridge University Press, Cambridge, 2011)
39. J. Rammer, H. Smith, Quantum field-theoretical methods in transport theory of metals. Rev. Mod. Phys. **58**, 323 (1986)
40. H. Haug, A. Jauho, *Quantum Kinetics in Transport and Optics of Semiconductors* (Springer, Berlin, 2008)
41. G. Stefanucci, R. van Leeuwen, *Nonequilibrium Many-Body Theory of Quantum Systems: A Modern Introduction* (Cambridge University Press, Cambridge, 2013)
42. G. Baym, L.P. Kadanoff, Conservation laws and correlation functions. Phys. Rev. **124**, 287 (1961)
43. J.K. Freericks, Quenching Bloch oscillations in a strongly correlated material: nonequilibrium dynamical mean-field theory. Phys. Rev. B **77**, 075109 (2008)
44. M. Eckstein, M. Kollar, Ph Werner, Interaction quench in the Hubbard model: relaxation of the spectral function and the optical conductivity. Phys. Rev. B **81**, 115131 (2010)
45. H. Brunner, P. J. van der Houwen, *The Numerical Solution of Volterra Equations*, Elsevier Science Ltd, (1986)
46. P. Lipavský, V. Špička, B. Velický, Generalized kadanoff-baym ansatz for deriving quantum transport equations. Phys. Rev. B **34**, 6933 (1986)
47. J. Li, C. Aron, G. Kotliar, J.E. Han, Electric-field-driven resistive switching in the dissipative hubbard model. Phys. Rev. Lett. **114**, 226403 (2015)
48. E. Pavarini, E. Koch, D. Vollhardt, A. Lichtenstein (eds.) *DMFT at 25: Infinite Dimensions*, Modeling and Simulation **4**, Verlag des Forschungszentrum Jülich, 2014, Jülich (2014)
49. W. Metzner, D. Vollhardt, Correlated lattice fermions in $d = \infty$ dimensions. Phys. Rev. Lett. **62**, 324 (1989)
50. A. Georges, Strongly correlated electron materials: dynamical mean-field theory and electronic structure. AIP Conference Proceedings **715**, 3 (2004)
51. A. Georges, G. Kotliar, W. Krauth, M.J. Rozenberg, Dynamical mean-field theory of strongly correlated fermion systems and the limit of infinite dimensions. Rev. Mod. Phys. **68**, 13 (1996)
52. P. Schmidt, H. Monien, *Nonequilibrium dynamical mean-field theory of a strongly correlated system*, arXiv:cond-mat/0202046 (unpublished)
53. J.K. Freericks, V.M. Turkowski, V. Zlatić, Nonequilibrium dynamical mean-field theory. Phys. Rev. Lett. **97**, 266408 (2006)
54. E. Müller-Hartmann, Correlated fermions on a lattice in high dimensions. Z. Phys. B **74**, 507 (1989)
55. E. Müller-Hartmann, The Hubbard model at high dimensions: some exact results and weak coupling theory. Z. Phys. B **76**, 211 (1989)
56. U. Brandt, C. Mielsch, Free energy of the Falicov-Kimball model in large dimensions. Z. Phys. B **82**, 37 (1991)
57. A. Georges, G. Kotliar, Hubbard model in infinite dimensions. Phys. Rev. B **45**, 6479 (1992)

58. V. Turkowski, J.K. Freericks, Nonlinear response of Bloch electrons in infinite dimensions. Phys. Rev. B **71**, 085104 (2005)
59. N. Tsuji, Ph Werner, Nonequilibrium dynamical mean-field theory based on weak-coupling perturbation expansions: application to dynamical symmetry breaking in the Hubbard model. Phys. Rev. B **88**, 165115 (2013)
60. E. Gull, A.J. Millis, A. Lichtenstein, A.N. Rubtsov, M. Troyer, Ph Werner, Continuous-time Monte Carlo methods for quantum impurity models. Rev. Mod. Phys. **83**, 349 (2011)
61. L. Mühlbacher, E. Rabani, Real-time path integral approach to nonequilibrium many-body quantum systems. Phys. Rev. Lett. **100**, 176403 (2008)
62. Ph Werner, T. Oka, A.J. Millis, Diagrammatic Monte Carlo simulation of non-equilibrium systems. Phys. Rev. B **79**, 035320 (2009)
63. M. Eckstein, M. Kollar, Ph Werner, Thermalization after an interaction quench in the hubbard model. Phys. Rev. Lett. **103**, 056403 (2009)
64. E. Canovi, Ph Werner, M. Eckstein, First order dynamical phase transitions. Phys. Rev. Lett. **113**, 265702 (2014)
65. G. Cohen, E. Gull, D.R. Reichman, A.J. Millis, Taming the dynamical sign problem in real-time evolution of quantum many-body problems. Phys. Rev. Lett. **115**, 266802 (2015)
66. Ph Werner, M. Eckstein, Phonon-enhanced relaxation and excitation in the Holstein-Hubbard model. Phys. Rev. B **88**, 165108 (2013)
67. D. Golez, M. Eckstein, Ph Werner, Dynamics of screening in photo-doped Mott insulators. Phys. Rev. B **92**, 195123 (2015)
68. H.U.R. Strand, M. Eckstein, Ph Werner, Nonequilibrium dynamical mean-field theory for bosonic lattice models. Phys. Rev. X **5**, 011038 (2015)
69. H.U.R. Strand, D. Golež, M. Eckstein, Ph Werner, Hund's coupling driven photo-carrier relaxation in the two-band Mott insulator. Phys. Rev. B **96**, 165104 (2017)
70. M. Eckstein, Ph Werner, Ultra-fast photo-carrier relaxation in Mott insulators with short-range spin correlations. Scientific Reports **6**, 21235 (2016)
71. M. Eckstein, Ph Werner, Nonequilibrium dynamical mean-field calculations based on the noncrossing approximation and its generalizations. Phys. Rev. B **82**, 115115 (2010)
72. E. Arrigoni, M. Knap, W. von der Linden, Nonequilibrium dynamical mean-field theory: an auxiliary quantum master equation approach. Phys. Rev. Lett. **110**, 086403 (2013)
73. Ch. Gramsch, K. Balzer, M. Eckstein, M. Kollar, Hamiltonian-based impurity solver for nonequilibrium dynamical mean-field theory. Phys. Rev. B **88**, 235106 (2013)
74. F.A. Wolf, I.P. McCulloch, U. Schollwöck, Solving nonequilibrium dynamical mean-field theory using matrix product states. Phys. Rev. B **90**, 235131 (2014)
75. F.A. Wolf, A. Go, I.P. McCulloch, A.J. Millis, U. Schollwöck, Imaginary-time matrix product state impurity solver for dynamical mean-field theory. Phys. Rev. X **5**, 041032 (2015)
76. D. Bauernfeind, M. Zingl, R. Triebl, M. Aichhorn, H.G. Evertz, Fork tensor-product states: efficient multiorbital real-time DMFT solver. Phys. Rev. X **7**, 031013 (2017)
77. X. Deng, J. Mravlje, R. Žitko, M. Ferrero, G. Kotliar, A. Georges, How bad metals turn good: spectroscopic signatures of resilient quasiparticles. Phys. Rev. Lett. **110**, 086401 (2013)
78. M. Dumm, D. Faltermeier, N. Drichko, M. Dressel, C. Mézière, P. Batail, Bandwidth-controlled Mott transition in κ–(BED–TTF)$_2$Cu[N(CN)$_2$]Br$_x$Cl$_{1-x}$: Optical studies of correlated carriers. Phys. Rev. B **79**, 195106 (2009)
79. P. Limelette, A. Georges, D. Jérome, P. Wzietek, P. Metcalf, J.M. Honig, Universality and Critical Behavior at the Mott Transition. Science **302**, 89 (2003)

80. M. Eckstein, Ph Werner, Thermalization of a pump-excited Mott insulator. Phys. Rev. B **84**, 035122 (2011)
81. N. Strohmaier, D. Greif, R. Jördens, L. Tarruell, H. Moritz, T. Esslinger, R. Sensarma, D. Pekker, E. Altman, E. Demler, Observation of elastic doublon decay in the fermi-hubbard model. Phys. Rev. Lett. **104**, 080401 (2010)
82. R. Sensarma, D. Pekker, E. Altman, E. Demler, N. Strohmaier, D. Greif, R. Jördens, L. Tarruell, H. Moritz, T. Esslinger, Lifetime of double occupancies in the Fermi-Hubbard model. Phys. Rev. B **82**, 224302 (2010)
83. S. Iwai, M. Ono, A. Maeda, H. Matsuzaki, H. Kishida, H. Okamoto, Y. Tokura, Ultrafast optical switching to a metallic state by photoinduced Mott transition in a halogen-bridged nickel-chain compound. Phys. Rev. Lett. **91**, 057401 (2003)
84. H. Okamoto, T. Miyagoe, K. Kobayashi, H. Uemura, H. Nishioka, H. Matsuzaki, A. Sawa, Y. Tokura, Ultrafast charge dynamics in photoexcited Nd_2CuO_4 and La_2CuO_4 cuprate compounds investigated by femtosecond absorption spectroscopy. Phys. Rev. B **82**, 060513(R) (2010)
85. M. Mitrano et al., Pressure-dependent relaxation in the photoexcited Mott insulator ET-F_2TCNQ: influence of hopping and correlations on quasiparticle recombination rates. Phys. Rev. Lett. **112**, 117801 (2014)
86. Z. Lenarčič, P. Prelovšek, Charge recombination in undoped cuprates. Phys. Rev. B **90**, 235136 (2014)
87. Z. Lenarčič, M. Eckstein, P. Prelovšek, Exciton recombination in one-dimensional organic Mott insulators. Phys. Rev. B **92**, 201104(R) (2015)
88. Ph Werner, K. Held, M. Eckstein, Role of impact ionization in the thermalization of photo-excited Mott insulators. Phys. Rev. B **90**, 235102 (2014)
89. E. Assmann, P. Blaha, R. Laskowski, K. Held, S. Okamoto, G. Sangiovanni, Oxide heterostructures for efficient solar cells. Phys. Rev. Lett. **110**, 078701 (2013)
90. Ph Werner, N. Tsuji, M. Eckstein, Nonthermal symmetry-broken states in the strongly interacting Hubbard model. Phys. Rev. B **86**, 205101 (2012)
91. R. Strack, D. Vollhardt, Dynamics of a hole in the t-J model with local disorder: exact results for high dimensions. Phys. Rev. B **46**, 13852 (1992)
92. M. Eckstein, Ph Werner, Ultrafast separation of photo-doped carriers in Mott antiferromagnets. Phys. Rev. Lett. **113**, 076405 (2014)
93. D. Golež, J. Bonča, M. Mierzejewski, L. Vidmar, Mechanism of ultrafast relaxation of a photo-carrier in antiferromagnetic spin background. Phys. Rev. B **89**, 165118 (2014)
94. M. Mierzejewski, L. Vidmar, J. Bonča, P. Prelovšsek, Nonequilibrium quantum dynamics of a charge carrier doped into a Mott insulator. Phys. Rev. Lett. **106**, 196401 (2011)
95. S. Dal Conte et al., Snapshots of the retarded interaction of charge carriers with ultrafast fluctuations in cuprates. Nature Phy. **11**, 421 (2015)
96. J.H. Mentink, M. Eckstein, Ultrafast quenching of the exchange interaction in a Mott insulator. Phys. Rev. Lett. **113**, 057201 (2014)
97. P. Beaud et al., A time-dependent order parameter for ultrafast photoinduced phase transitions. Nat. Mater. **923**, 13 (2014)
98. M. Eckstein, Ph Werner, Photo-induced states in a Mott insulator. Phys. Rev. Lett. **110**, 126401 (2013)
99. Sh Sayyad, M. Eckstein, Slow down of the electronic relaxation close to the Mott transition. Phys. Rev. Lett. **117**, 096403 (2016)
100. S. Florens, A. Georges, Quantum impurity solvers using a slave rotor representation. Phys. Rev. B **66**, 165111 (2002)
101. M. Bukov, L. D'Alessio, A. Polkovnikov, Universal high-frequency behavior of periodically driven systems: from dynamical stabilization to Floquet engineering Adv. Physics **64**, 139 (2015)
102. Y. Murakami, N. Tsuji, M. Eckstein, Ph. Werner, *Nonequilibrium steady states and transient dynamics of superconductors under phonon driving*, arXiv:1702.02942

103. M. Sandri, M. Fabrizio, Nonequilibrium gap collapse near a first-order Mott transition. Phys. Rev. B **91**, 115102 (2015)
104. G. Mazza, A. Georges, *Non-equilibrium Superconductivity in driven alkali-doped fullerides*, arxiv:1702.04675
105. Ph Werner, H. Strand, S. Hoshino, M. Eckstein, Ultrafast switching of composite order in A3C60. Phys. Rev. B **95**, 195405 (2017)
106. D. Golez, L. Boehnke, H. Strand, M. Eckstein, P. Werner, Nonequilibrium GW+EDMFT: antiscreening and inverted populations from nonlocal correlations. Phys. Rev. Lett. **118**, 246402 (2017)

Chapter 4
Master Equations Versus Keldysh Green's Functions for Correlated Quantum Systems Out of Equilibrium

Enrico Arrigoni and Antonius Dorda

Abstract The goal of these lecture notes is to illustrate connections between two widely used, but often separately adopted approaches to deal with quantum systems out of equilibrium, namely quantum master equations and nonequilibrium Green's functions. For the paradigmatic case of the Anderson impurity model out of equilibrium we elaborate on these connections and map its description from one approach to the other. At the end of this chapter, we will show how the "best of the two worlds" can be combined to obtain a highly accurate solution of this model, which resolves the nonequilibrium Kondo physics down to temperatures well below the Kondo scale. As a training course, these lectures devote a large portion to an introduction to the Lindblad quantum master equation based on standard treatments, as well as methods to solve this equation. For nonequilibrium Green's functions, which are discussed in the first part of the course, we only provide a summary of the most important aspects necessary to address the topics of the present chapter. The relevant aspects of these two topics are presented in a self-contained manner so that a background in equilibrium many-body physics is sufficient to follow these notes.

4.1 Introduction

The problem we address in these lectures consists of a small correlated central system in which particles interact with each other, connected to external noninteracting infinite reservoirs (leads), see Fig. 4.1. We focus here to the case of a purely fermionic model, although many ideas can be easily extended to more general problems including, e.g., electron-phonon interactions, photons, etc. We are typically interested in the case of two leads with different chemical potentials and/or temperatures (see, e.g. [1, 2]). Thus, a particle current flows from the lead with larger chemical poten-

E. Arrigoni (✉) · A. Dorda
Graz University of Technology, 8010 Graz, Austria
e-mail: arrigoni@tugraz.at

A. Dorda
e-mail: antonius.dorda@gmx.at

R. Citro and F. Mancini (eds.), *Out-of-Equilibrium Physics of Correlated Electron Systems*, Springer Series in Solid-State Sciences 191,
https://doi.org/10.1007/978-3-319-94956-7_4

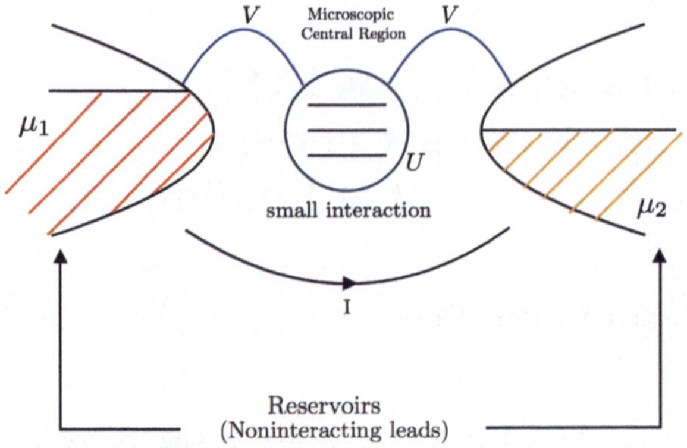

Fig. 4.1 Schematic illustration of the system of interest: A small interacting central system is connected to two leads with different chemical potentials and/or temperatures. The leads are infinite, so that a stationary current flows from one lead to the other in the steady state

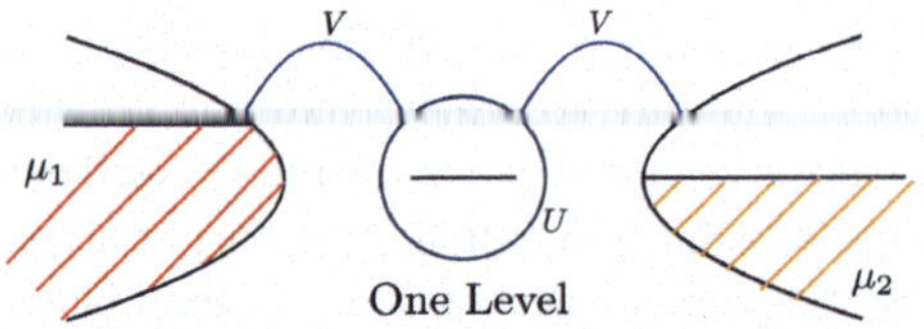

Fig. 4.2 Special case of the system depicted in Fig. 4.1: The single impurity Anderson Model out of equilibrium. The central system consists of a single spin-degenerate level, with on-site Hubbard interaction U. The coupling to the leads is provided by the hopping V

tial through the central system to the other lead, and, since the leads are infinite they provide the dissipation necessary to reach a stationary state. As a paradigmatic example, on which we will focus in the last part of this lecture, we consider the special case of a central system consisting of a single interacting spin-degenerate level with an onsite Hubbard interaction (Fig. 4.2), the single impurity Anderson model (SIAM) [3–7] out of equilibrium. This model is, on the one hand, interesting *per se* as a simple description of transport across quantum dots or small molecules and for understanding the Kondo effect, and on the other hand, constitutes the "bottleneck" problem in the self-consistent cycle within nonequilibrium dynamical mean field theory (DMFT) [8–24], see also previous chapters in this book. Therefore, an accurate solution of impurity models is of great interest and importance.

While we will restrict mainly to the steady state, other related situations can be treated with the approaches presented here and similar ones. For example, one can include a periodic driving within a Floquet approach, or study quantum quenches in which one is interested in the real time dynamics after a sudden change of parameters (see Fig. 4.3).

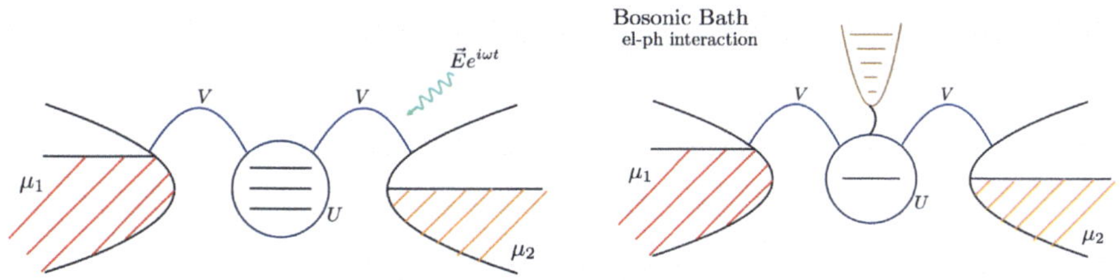

Fig. 4.3 Possible extensions: Time-dependent situations such as quantum quenches [15, 25], i.e. sudden change of parameters, or periodically driven systems [11, 26–29] (left); or coupling to a bosonic bath to account e.g. for electron phonon interaction [30–35] (right)

Limits in which the impurity problem can be solved

Numerous approaches have been developed in the past decades to address impurity problems. For the case of a single reservoir, the steady state corresponds to thermodynamic equilibrium. Dedicated methods such as numerical renormalization group (NRG) were developed in order to resolve the challenging and exponentially small low-energy physics of the model. As a result, the so-called Kondo effect (in equilibrium) is nowadays well understood [7, 36–38]. For the nonequilibrium case a number of numerical approaches have been suggested that are valid in specific limits, but a "full solution" is not yet available. A discussion of all these approaches is beyond the scope of these lectures.[1] Instead, we present in Sect. 4.9 the so-called Auxiliary Master Equation Approach (AMEA) a non-perturbative approach devised and developed in our group in recent years, which is based on a combination of quantum master equation with nonequilibrium Green's functions.

To start with, let consider situations in which models of Figs. 4.1 and 4.2 are exactly solvable. Two trivial cases are (i) the noninteracting case $U = 0$, where one can use nonequilibrium Green's functions, and the decoupled case $V = 0$, for which one can explicitly carry out an exact diagonalization of the many-body Fock space of the small central system. But there is another less trivial limit in which an exact solution is available, the so-called *Markovian Limit*. This is the case when the response of the reservoir is instantaneous, i.e. without memory effects. In this limit, the reservoir's degrees of freedom can be eliminated and the dynamics of the reduced density matrix of the central system is exactly described by the so-called *Lindblad Equation*. Since the central system is small this can again be solved by *exact diagonalization in the space of many-body density matrices*. This will be discussed in detail in Sect. 4.9.1.2.

Outline

The Lindblad equation is the main topic of the first part of the present lectures. More specifically, we will:

[1]For a non-comprehensive list see, e.g., [21, 39–64].

(1) Provide a derivation of the Lindblad Equation, first heuristically in Sect. 4.4.1 and then rigorously in Sect. 4.4.4.
(2) Discuss under which condition a reservoir can be considered as Markovian (Sect. 4.4.4). We will specify this explicitly in terms of the parameters of the microscopic model. As anticipated, we will mainly concentrate on fermionic models.
(3) Present some approaches to solve the many-body Lindblad equation for the noninteracting and the interacting case. In particular, we shall present the so-called superfermion representation [65] (Sect. 4.5), in which the space of density operators for the open system is replaced by a "superspace" of state vectors acting on twice as many single-particle levels (see also [66, 67]). In this formalism, the Lindblad equation acquires a structure like the Schrödinger equation, with which many of us are more familiar.
In the noninteracting case, this linear operator problem can be solved by equations-of-motion techniques, leading to an analytic expression for the steady-state Green's functions, see Sect. 4.9.1.1. In the interacting case we will discuss the solution via exact diagonalization in Sect. 4.9.1.2.

Master Equation Approaches

Unfortunately, it turns out that the Markovian approximation is unrealistic for interesting fermionic models. As we will see, a Markovian reservoir must have both a constant density of states as well as an infinite temperature T and/or chemical potential(s) μ.[2] While these two conditions appear quite restrictive and unphysical, a possible solution is to introduce an intermediate auxiliary buffer zone (mesoreservoir) between the Lindblad couplings and the central system (Sect. 4.8.1, see, e.g. [65, 69, 70] and Fig. 4.4). The buffer zone consists of N_B isolated discrete sites (bath levels), each one coupled to a reservoir with a constant density of states that is either completely filled $\mu = +\infty$ or completely empty $\mu = -\infty$. Therefore, these reservoirs fulfill the Markovian condition and the system can be exactly mapped onto a Lindblad equation. With properly chosen parameters and for large enough N_B, the buffer layer plus Markovian reservoirs exactly describe an *arbitrary non-Markovian* reservoir of noninteracting fermions coupled to the central system [71].[3]

Here comes the connection with nonequilibrium Green's functions[4]: The model as depicted in the lower part of Fig. 4.4 can, on the one hand, be seen as a an open system (the central system plus the buffer layer) whose dynamics is exactly controlled by the Lindblad equation, and on the other hand, consists of a closed system with an infinite number of fermionic levels, that can be (approximately) treated by nonequilibrium Green's functions. For the case of a noninteracting central system, an exact solution is obviously available in both cases. This is shown in Sect. 4.9.1.1, where we will derive analytic expressions for the steady state Green's functions in the noninteracting case.

---

[2]See also [68].

[3]For related discussions about non-Markovianity and open quantum systems, see e.g. [72–76].

[4]For similar discussions see, e.g. [52, 77].

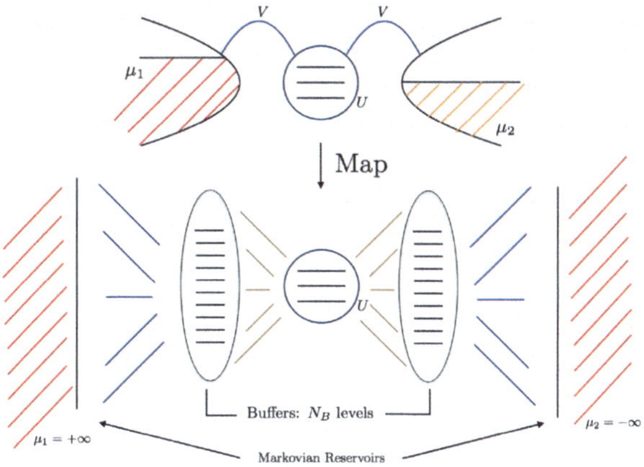

Fig. 4.4 Buffer layer approach: Mapping the original problem onto a system with a finite number of levels, connected in turn to Markovian environments. With appropriate choice of the parameters, the mapping becomes exact for $N_B \to \infty$

The problem we want to address, however, includes an interaction in the central system, which makes an exact solution by Green's function methods impossible. On the other hand, the Lindblad equation for the open system can, in principle, be solved exactly by approaches addressing the full many body space of density matrices, provided N_B is small enough. This will be discussed in Sect. 4.9.1.2. Unfortunately, the buffer layer representation discussed above is limited by the fact that an accurate description of the original system requires quite large values N_B, especially at low temperatures where the Fermi function is sharp. Consequently, the many-body Hilbert space is too large and the treatment of a correlated problem becomes prohibitive.

In the last part of these lectures, Sect. 4.9, we will illustrate how the efficiency of this buffer layer approach can be significantly improved by allowing for more general Lindblad couplings [21, 63, 71, 78], which are determined through an optimization procedure aiming at fitting the so-called bath hybridization function. For the case of the nonequilibrium Anderson impurity model, Fig. 4.2, already modest values of N_B ($N_B \lesssim 8$), which can be treated by Krylov-space schemes [63], are sufficient to resolve the nonequilibrium behavior of the Kondo resonance. Larger values of N_B ($N_B \lesssim 20$) can be addressed by matrix-product states [79–82] which allows to resolve the Kondo peak at temperatures below the Kondo scale with an accuracy that, in equilibrium, becomes comparable with NRG up to intermediate values of the interaction U [78].[5]

4.2 Master Equations

Besides quantum problems, master equations are a central object in classical physics in the context of stochastic processes. Examples are for instance Brownian motion

[5]More specifically, in [78] we resolved temperatures down to $T/T_K = 0.25$ for $T_K = 0.2\Gamma$.

or any other subsystem coupled to a heat bath/environment. In such cases, when the dynamics of a system is non-deterministic it is convenient to describe its state by a probability density. For the case of stochastic processes which fulfill the *Markov* property, i.e. which have a very short memory kernel and only depend on the present state of the system, a master equation is applicable. For a thorough introduction we refer e.g. to [83–85]. Here, we will follow the treatment by Schaller [84].

Let us consider here a discrete set of system states labeled by k and assign each state a probability P_k. The temporal evolution of these probabilities is governed by the rates $T_{kl} \geq 0$ for a transition from state l to state k, and is described by the master equation:

$$\boxed{\frac{dP_k}{dt} = \sum_l [T_{kl}P_l - T_{lk}P_k]} \tag{4.1}$$

In order for the P_k to be interpreted as probabilities, at each time they have to obey two properties: (i) Conservation of total probability $\sum_k P_k = 1$, and (ii) semipositivity $P_k \geq 0 \ \forall k$. Assuming that (i) and (ii) are fulfilled at some initial time, the master equation must guarantee that these properties are preserved.
(i) can be proven as follows:

$$\sum_k \frac{dP_k}{dt} = \sum_{kl} (T_{kl}P_l - T_{lk}P_k) = \sum_{kl} (T_{lk}P_k - T_{lk}P_k) = 0\,. \tag{4.2}$$

For (ii) one can argue in the following way. Assume that for a certain k^*, the corresponding P_{k^*} becomes zero at a certain time. Then

$$\left.\frac{dP_{k^*}}{dt}\right|_{P_{k^*}=0} = +\sum_l T_{k^*l}P_l \geq 0\,. \tag{4.3}$$

Therefore, P_{k^*} cannot become negative.

Example Consider the temporal dynamics of a two level system with transition rate T_{10} from state $k = 0$ to state $k = 1$ and rate T_{01} for the inverse process. The master equation (4.1) is in matrix form then given by

$$\frac{d}{dt}\begin{pmatrix} P_0 \\ P_1 \end{pmatrix} = \begin{pmatrix} -T_{10} & +T_{01} \\ +T_{10} & -T_{01} \end{pmatrix}\begin{pmatrix} P_0 \\ P_1 \end{pmatrix}. \tag{4.4}$$

The stationary (steady-state) solution P_i^∞ is obtained by setting the left-hand side to zero, yielding

$$\begin{aligned} P_0^\infty &= T_{01}/\,(T_{01} + T_{10})\,, \\ P_1^\infty &= T_{10}/\,(T_{01} + T_{10})\,. \end{aligned} \tag{4.5}$$

The two eigenvalues of the matrix on the right-hand side of (4.4) are 0 and $-\lambda = -T_{01} - T_{10}$. The former corresponds to the stationary solution and λ determines the decay rate in the time evolution. The full time-dependent solution of (4.4) is easily seen to be

$$\begin{aligned} P_0(t) &= P_0(0)e^{-\lambda t} + P_0^{\infty}\left(1 - e^{-\lambda t}\right), \\ P_1(t) &= P_1(0)e^{-\lambda t} + P_1^{\infty}\left(1 - e^{-\lambda t}\right). \end{aligned} \tag{4.6}$$

4.3 Density Matrix

Open quantum systems consist of a microscopic quantum mechanical central system of interest which couples, possibly weakly, with an environment. Due to the entanglement with the environment, the properties of the central system cannot be described by a quantum state alone, but rather require the concept of reduced density matrix. The same concept is also needed if the quantum state of the central system is not known exactly due to a statistical uncertainty. A general mixed system state can be characterized by an ensemble of states $\{|\Phi_i\rangle\}$ which are realized with probability P_i. Here $\sum_i P_i = 1$ and the states are normalized but not necessarily orthogonal. Such a mixed quantum state is conveniently described in terms of the density matrix (or density operator)

$$\rho = \sum_i P_i |\Phi_i\rangle\langle\Phi_i| \,. \tag{4.7}$$

The expectation value of an operator A for the system is then given by

$$\begin{aligned} \langle A\rangle &= \sum_i P_i \langle\Phi_i|A|\Phi_i\rangle \\ &= \sum_{i,n} P_i \langle\Phi_i|n\rangle\langle n|A|\Phi_i\rangle \\ &= \sum_n \langle n|A \underbrace{\sum_i P_i|\Phi_i\rangle\langle\Phi_i|\, n\rangle}_{\text{Density matrix}} \\ &= \mathrm{Tr} A\rho \,. \end{aligned} \tag{4.8}$$

ρ must fulfill the following properties:

$\langle 1\rangle = 1 \Rightarrow \mathrm{Tr}\rho = 1$	Normalization
$\rho = \rho^{\dagger}$	Hermiticity
$\langle\psi|\rho|\psi\rangle = \sum_i P_i |\langle\psi|\Phi_i\rangle|^2 \geq 0 \quad \forall|\psi\rangle$	(Semi) positivity: $\rho \geq 0$

If the system is characterized by a single quantum mechanical state with probability 1, the density matrix describes a so-called pure state for which

$$\rho = |\Phi_i\rangle\langle\Phi_i| \Rightarrow \rho^2 = \rho . \tag{4.9}$$

On the other hand, for a general non-pure (mixed) state $\rho = \sum_n P_n |\psi_n\rangle\langle\psi_n|$ expanded in its eigenbasis[6] one finds

$$\begin{aligned} \mathrm{Tr}\rho^2 &= \sum_{m,n,k} \langle\psi_m|\psi_n\rangle\langle\psi_n|\psi_k\rangle\langle\psi_k|\psi_m\rangle P_n P_k \\ &= \sum_n P_n^2 \leq 1 . \end{aligned} \tag{4.10}$$

Therefore, a system is in a pure state if and only if

$$\mathrm{Tr}\rho^2 = 1 \tag{4.11}$$

so that $\mathrm{Tr}\rho^2$ is a measure for the degree of purity of the state [86].

4.3.1 Time Dependence

The time evolution of the density matrix ρ for a *closed* quantum system is determined by the Liouville von Neumann Equation[7]:

$$\begin{aligned} \dot{\rho} &= \sum_i P_i \left(|\dot{\Phi}_i\rangle\langle\Phi_i| + |\Phi_i\rangle\langle\dot{\Phi}_i|\right) \\ &= -i[H, \rho] , \end{aligned} \tag{4.12}$$

which can be easily obtained by applying the Schrödinger equation for $|\dot{\Phi}_i\rangle$, and using $\dot{P}_i = 0$. Notice that (4.12) is similar to the Heisenberg time evolution for an operator, $\dot{A} = +i[H, A]$, however, the sign is opposite.

It is easy to verify that (4.12) preserves normalization, Hermiticity and semipositivity of the density matrix:

$$\mathrm{Tr}\rho = 1 \qquad \rho = \rho^\dagger \qquad \rho \geq 0 , \tag{4.13}$$

An important example for a nonunitary evolution is a measure operation. Let us consider the spectral representation of a generic operator A,[8]

---

[6]In contrast to the $|\Phi_i\rangle$ in (4.7) the $|\psi_n\rangle$ are orthogonal to each other.

[7]From now on we will adopt units in which $\hbar = 1$.

[8]In these lectures, we will not explicitly mark operators with a hat "^" except when there is a risk of confusion.

$$\begin{aligned} A &= \sum_n a_n \hat{P}_n, \\ \hat{P}_n &\equiv |a_n\rangle\langle a_n|, \end{aligned} \tag{4.14}$$

with $\hat{P}_n$ the projection operators onto the eigenstates $|a_n\rangle$, of A, i.e. $A|a_n\rangle = a_n|a_n\rangle$. Quantum mechanics tells us that if one measures A on a pure state $|\Phi_i\rangle$, the value a_n is obtained with probability $P_n = |\hat{P}_n|\Phi_i\rangle|^2$, and the state collapses to

$$|\Phi_i\rangle \to \frac{\hat{P}_n|\Phi_i\rangle}{\sqrt{P_n}}. \tag{4.15}$$

Therefore, when starting from $\rho = |\Phi_i\rangle\langle\Phi_i|$ and performing a measure without looking at the result one gets an ensemble of states with the probabilities P_n, i.e.

$$\begin{aligned} \rho \xrightarrow{\text{Measure}} & \sum_n P_n \left(\frac{\hat{P}_n|\Phi_i\rangle}{\sqrt{P_n}}\right)\left(\frac{\langle\Phi_i|\hat{P}_n}{\sqrt{P_n}}\right) \\ = & \sum_n \hat{P}_n|\Phi_i\rangle\langle\Phi_i|\hat{P}_n = \sum_n \hat{P}_n \rho \hat{P}_n . \end{aligned} \tag{4.16}$$

Clearly, the last line of (4.16), the *von Neumann measure*, holds also in the case in which one starts from a mixed state ρ. Also in the case of a von Neumann measure, the properties (4.13) are preserved.

Unitary evolution and von Neumann measure are two examples of *quantum operations*, i.e. linear time evolutions for the density matrix.

Example The density matrix of a spin 1/2 system, or any other two-state quantum system, can be represented in terms of the so-called Bloch sphere. The density matrix for such a system can be expressed in terms of the identity I and the Pauli matrices $\boldsymbol{\sigma}$

$$\rho = \frac{1}{2}(I + \boldsymbol{\alpha} \cdot \boldsymbol{\sigma}), \tag{4.17}$$

with $\boldsymbol{\alpha}$ an appropriate vector with real coefficients. From

$$\begin{aligned} \text{Tr}\rho^2 &= \frac{1}{4}\,\text{Tr}\left(I + 2\boldsymbol{\alpha}\cdot\boldsymbol{\sigma} + (\boldsymbol{\alpha}\cdot\boldsymbol{\sigma})^2\right) \\ &= \frac{1}{4}(2 + 2|\alpha|^2)\,, \end{aligned} \tag{4.18}$$

using $\text{Tr}\sigma_i\sigma_j = 2\delta_{ij}$, one finds that $|\boldsymbol{\alpha}| = 1$ describes a pure state, while a mixed state has $|\boldsymbol{\alpha}| < 1$.

4.3.2 Reduced Density Matrix

Open quantum systems consist of a microscopic system embedded within a reservoir, see also Fig. 4.5. In general, one is not interested in the properties of the reservoir itself, however, the latter affect the dynamics of the system. Ideally, one would like to eliminate the degrees of freedom of the reservoir and obtain an effective description for the system alone. Due to the entanglement with the reservoir, the system's quantum mechanical state must be formulated in terms of the so-called reduced density matrix, which, quite generally, describes a mixed state.

The combined Hilbert space of the so-called "universe" (= system + reservoir) is given by the *tensor product space* of the system and the reservoir Hilbert spaces $\mathcal{H}_S$ and $\mathcal{H}_R$:

$$\mathcal{H}_U = \mathcal{H}_S \otimes \mathcal{H}_R. \tag{4.19}$$

A basis of $\mathcal{H}_U$ is $\{|S_i\rangle \otimes |R_\alpha\rangle\}$, where $\{|S_i\rangle\}$ is a basis of $\mathcal{H}_S$ and $\{|R_\alpha\rangle\}$ a basis of $\mathcal{H}_R$. For simplicity, we will use alternative equivalent notations

$$|S_i\rangle \otimes |R_\alpha\rangle \leftrightarrow |S_i\rangle|R_\alpha\rangle \leftrightarrow |S_i, R_\alpha\rangle \,, \tag{4.20}$$

and for the bra counterparts

$$\langle S_i| \otimes \langle R_\alpha| \leftrightarrow \langle S_i|\langle R_\alpha| \leftrightarrow \langle S_i|\langle R_\alpha| \,.$$

Let us recall the following important properties of the tensor product:

- Distributivity:

$$\begin{aligned}(|a\rangle + |b\rangle) \otimes |c\rangle &= |a\rangle \otimes |c\rangle + |b\rangle \otimes |c\rangle \\ &= |a, c\rangle + |b, c\rangle \end{aligned} \tag{4.21}$$

- Operators act only on states in their corresponding subspace:

$$(A \otimes B)|x\rangle \otimes |y\rangle = (A|x\rangle) \otimes (B|y\rangle) \tag{4.22}$$

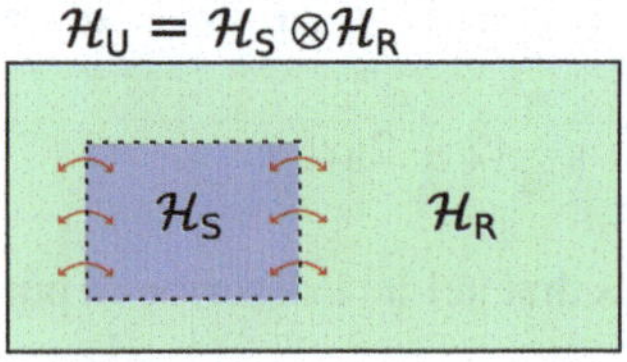

Fig. 4.5 Open system embedded into a reservoir. The Hilbert space for the full "universe" is given by $\mathcal{H}_U = \mathcal{H}_S \otimes \mathcal{H}_R$, for which a pure state description in terms of a wave function is possible. Due to particle/energy exchange between the system and the reservoir this is not true for $\mathcal{H}_S$ and $\mathcal{H}_R$ separately, which requires a description in terms of density matrices

- Scalar Product:

$$(\langle a| \otimes \langle b|) \otimes (|x\rangle \otimes |y\rangle) = \langle a|x\rangle\langle b|y\rangle \tag{4.23}$$

As noted above, the set of product states (4.20) provides a complete basis for $\mathcal{H}_U$. However, a generic state in $\mathcal{H}_U$ will not factorize in simple product states in terms of the system and the reservoir separately. One then speaks of *entangled states*. As an example, let us consider two basis states for the system and the reservoir each, $|S_i\rangle$ and $|R_\alpha\rangle$, $i, \alpha = 1, 2$ and the following two states:

$$\begin{aligned} |\psi_P\rangle &= |S_1\rangle|R_1\rangle + |S_2\rangle|R_1\rangle \\ &= (|S_1\rangle + |S_2\rangle)|R_1\rangle, \end{aligned} \tag{4.24}$$

and

$$\begin{aligned} |\psi_E\rangle &= |S_1\rangle|R_1\rangle + |S_2\rangle|R_2\rangle \\ &\neq |a\rangle|b\rangle. \end{aligned} \tag{4.25}$$

While $|\psi_p\rangle$ is a product state, the latter is not.

From the properties (4.22) and (4.23) one can compute the trace of a tensor product operator $\hat{O} = \hat{A} \otimes \hat{B}$ as

$$\begin{aligned} \mathrm{Tr}\hat{O} &= \sum_{i,\alpha} \langle S_i, R_\alpha|\hat{O}|S_i, R_\alpha\rangle \\ &= \sum_{i,\alpha} \langle S_i|\hat{A}|S_i\rangle\langle R_\alpha|\hat{B}|R_\alpha\rangle \\ &= \underbrace{\mathrm{Tr}\hat{A}}_{\mathrm{Tr}_S}\ \underbrace{\mathrm{Tr}\hat{B}}_{\mathrm{Tr}_R}\ . \end{aligned} \tag{4.26}$$

In the last line we introduced the partial traces over either system or reservoir states, defined as:

$$\begin{aligned} \mathrm{Tr}_S\, \hat{O} &= \sum_i \langle S_i|\hat{O}|S_i\rangle, \\ \mathrm{Tr}_R\, \hat{O} &= \sum_\alpha \langle R_\alpha|\hat{O}|R_\alpha\rangle. \end{aligned} \tag{4.27}$$

Let us consider for instance the operator $\hat{O} = |S\rangle\langle S'| \otimes |R\rangle\langle R'|$ and evaluate

$$\begin{aligned} \mathrm{Tr}_R\hat{O} &= |S\rangle\langle S'| \sum_\alpha \langle R_\alpha|R\rangle\langle R'|R_\alpha\rangle \\ &= |S\rangle\langle S'|\, \mathrm{Tr}_R|R\rangle\langle R'|. \end{aligned} \tag{4.28}$$

From this one sees that the partial trace Tr_R produces an operator acting in $\mathcal{H}_S$ alone. An explicit expression for the partial trace of an arbitrary operator $\hat{O}$ expanded in the product basis (4.20)

$$\hat{O} = \sum_{i,\alpha,j,\beta} O_{i,\alpha,j,\beta} |S_i\rangle|R_\alpha\rangle\langle S_j|\langle R_\beta|$$

can be readily obtained as

$$\mathrm{Tr}_R\hat{O} = \sum_{i,j}\left(\sum_\alpha O_{i,\alpha,j,\alpha}\right)|S_i\rangle\langle S_j|. \tag{4.29}$$

If we are interested in observables of the system only we can restrict to the system's *reduced density matrix*

$$\rho_S = \mathrm{Tr}_R\rho, \tag{4.30}$$

which is obtained as the partial trace of the density matrix ρ of the universe over the reservoir degrees of freedom. Indeed, the expectation value of an arbitrary operator $A \otimes I$ acting on the system only can be expressed as

$$\begin{aligned}\langle A \otimes I\rangle &= \mathrm{Tr}\,\{(A \otimes I)\rho\}\\ &= \sum_{i,\alpha}\left(\langle S_i|A \otimes \langle R_\alpha|\right)\rho\left(|S_i\rangle \otimes |R_\alpha\rangle\right)\\ &= \sum_i \langle S_i|A\sum_\alpha\langle R_\alpha|\rho|R_\alpha\rangle|S_i\rangle\\ &= \sum_i\langle S_i|A(\mathrm{Tr}_R\rho)|S_i\rangle\\ &= \mathrm{Tr}_S A\rho_S,\end{aligned} \tag{4.31}$$

which is valid for any system operator A. The reduced density matrix, thus, contains all the necessary information to compute system properties. Quite generally, for the universe one can assume that the density matrix is represented by a pure state $\rho = |\psi\rangle\langle\psi|$. On the contrary, the reduced system density matrix $\rho_S = \mathrm{Tr}_R|\psi\rangle\langle\psi|$ only describes a pure state if the universe wave function is a product state: $|\psi\rangle = |R\rangle|S\rangle$. In the general case, when $|\psi\rangle$ is entangled, ρ_S describes a mixed state with $\mathrm{Tr}\rho_S^2 < 1$.

On the other hand, for every given system density matrix ρ_S one can always construct a "sufficiently large" universe $\mathcal{H}_U = \mathcal{H}_S \otimes \mathcal{H}_R$ such that

$$\rho_S = \mathrm{Tr}_R|\psi\rangle\langle\psi|\,, \tag{4.32}$$

and with $|\psi\rangle$ a pure state. This procedure is termed *purification*. For example, suppose we have $\rho_S = \sum_{n=1}^N P_n|\Phi_n\rangle\langle\Phi_n|$. In this case one needs a reservoir with an

N-dimensional orthonormal basis set $\{|R_n\rangle\}$. A universe wave function $|\psi\rangle$ satisfying (4.32) is then, for instance, given by

$$|\psi\rangle = \sum_n \sqrt{P_n}|\Phi_n\rangle \otimes |R_n\rangle \; .$$

Proof

$$\begin{aligned}\mathrm{Tr}_R|\psi\rangle\langle\psi| &= \sum_{n,m} \sqrt{P_n}\sqrt{P_m}|\Phi_n\rangle\langle\Phi_m| \underbrace{\mathrm{Tr}_R|R_n\rangle\langle R_m|}_{\delta_{n,m}} \; . \\ &= \rho_S\end{aligned}$$

The above examples show that there can be two situations in which the quantum mechanical state $|\psi\rangle$ is not sufficient to describe a microscopic system and one needs a density matrix:

1. The exact state is not known, only its statistical distribution.
2. The system is entangled with a reservoir.

In the rest of these lectures we will be interested in the second case.

4.4 Lindblad Equation

As discussed in the previous section, the reduced density matrix contains all possible information about a microscopic system even if it is in contact (entangled) with a large reservoir. This is obviously a big advantage since one has to deal with a much smaller Hilbert space, without caring about the much larger reservoir. However, computing the time evolution of the reduced density matrix is again a prohibitive task. Whenever there is a coupling between system and environment, ρ_S does not evolve according to the Liouville equation (4.12). To find its time evolution one should first evolve the density matrix of the universe ρ, which follows the Liouville equation, and then carry out the partial trace (4.29). The intermediate step, thus, involves again addressing the full universe Hilbert space. It would be useful if, under some conditions, one could work in the restricted subspace of the system reduced density matrices including the action of the reservoir in some effective way. In this section, we are going to show that within the so-called *Markovian condition* one can indeed formulate the time evolution of the reservoir within a closed time evolution equation for ρ_S, the *Lindblad equation*. In Sect. 4.4.4 we will present a microscopic derivation of the Lindblad equation in the so-called strong-coupling limit, and discuss under which conditions, in terms of the parameters of the microscopic model, this equation provides an exact description of the effects of the environment. A microscopic derivation, as well as a derivation obtained by the conventional Markovian assumption based on so-called Kraus operators can be found in several textbooks, see, e.g. [83–85, 87, 88]. Here we will roughly follow the treatments of [84, 88].

But before becoming rigorous, we first present a heuristic derivation based on the master equation discussed in Sect. 4.2.

4.4.1 Heuristic Derivation

A situation described in Sect. 4.3 in which one has a set of quantum-mechanical states $|k\rangle$ occupied with probabilities P_k (also called populations) is described by a density matrix with diagonal elements $\rho_{k,k} = P_k$. As a consequence, the Master equation (4.1) can be rewritten as

$$\left.\frac{d\rho_{kk}}{dt}\right|_{Master} = \sum_l (T_{kl}\rho_{ll} - T_{lk}\rho_{kk}), \tag{4.33}$$

The transitions between different states in (4.33) can be expressed in terms of *jump operators*

$$\hat{J}_{kl} = |k\rangle\langle l|. \tag{4.34}$$

This allows to express (4.33) in operator form

$$\left.\frac{d\hat{\rho}}{dt}\right|_{Master} = \sum_{l,k} (T_{kl}\hat{J}_{kl}\hat{\rho}\hat{J}^{\dagger}_{kl} - T_{lk}\hat{J}_{kk}\hat{\rho}). \tag{4.35}$$

In order to have an expression which is quadratic in $\hat{J}$, we rewrite $\hat{J}_{kk} = \hat{J}_{kn}\hat{J}_{nk} = \hat{J}^{\dagger}_{nk}\hat{J}_{nk}$ with arbitrary n. Accordingly, the term $\hat{J}_{kk}\hat{\rho}$ can be written in several forms, for instance $\hat{J}^{\dagger}_{lk}\hat{J}_{lk}\hat{\rho}$ or $\hat{\rho}\hat{J}^{\dagger}_{lk}\hat{J}_{lk}$. While these give the same result for the diagonal terms (4.33), different results are obtained for the nondiagonal terms. We here choose[9] the symmetrized form

$$\hat{J}_{kk}\hat{\rho} \to \frac{1}{2}\left\{\hat{J}^{\dagger}_{lk}\hat{J}_{lk}, \hat{\rho}\right\}, \tag{4.36}$$

leading to

$$\left.\frac{d\hat{\rho}}{dt}\right|_{Master} = \sum_{l,k} T_{kl}\left(\hat{J}_{kl}\hat{\rho}\hat{J}^{\dagger}_{kl} - \frac{1}{2}\left\{\hat{J}^{\dagger}_{kl}\hat{J}_{kl}, \hat{\rho}\right\}\right). \tag{4.37}$$

We now replace the jump operators by arbitrary operators

$$J_{kl} \to \bar{S}_n$$

[9]Remember, this is just a non-rigorous derivation.

with corresponding coefficients

$$T_{kl} \rightarrow \bar{\gamma}_n$$

and omit the "hats" for the sake of readability. In addition to the master equation (4.33), which describes changes in population of the states, one has to include the Liouville von Neumann contribution (4.12) originating from the internal Hamiltonian dynamics, which leads to

$$\frac{d\rho}{dt} = -i\,[H,\rho] + \sum_n \bar{\gamma}_n \left(\bar{S}_n \rho \bar{S}_n^\dagger - \frac{1}{2}\left\{\bar{S}_n^\dagger \bar{S}_n, \rho\right\}\right). \tag{4.38}$$

This is the *Lindblad equation* in diagonal form.

The positivity of probabilities is ensured by using nonnegative coefficients $\bar{\gamma}_n$. This allows them to be absorbed into the definition of the $\bar{S}_n$ operators

$$\Gamma_n = \sqrt{\bar{\gamma}_n}\bar{S}_n, \tag{4.39}$$

so that (4.38) can be written as

$$\frac{d\rho}{dt} = -i\,[H,\rho] + \sum_n \left(\Gamma_n \rho \Gamma_n^\dagger - \frac{1}{2}\left\{\Gamma_n^\dagger \Gamma_n, \rho\right\}\right). \tag{4.40}$$

Besides the diagonal form (4.40), also a non-diagonal one is often adopted:

$$\frac{d\rho}{dt} = -i\,[H,\rho] + \sum_{\alpha\beta} \gamma_{\alpha\beta} \left(S_\beta \rho S_\alpha^\dagger - \frac{1}{2}\left\{S_\alpha^\dagger S_\beta, \rho\right\}\right), \tag{4.41}$$

where the coefficient matrix $\gamma_{\alpha\beta}$ is Hermitian and semi-positive definite. As can be easily checked, the two forms are linked by the eigen decomposition $\gamma_{\alpha\beta} = \sum_n U^\dagger_{\alpha,n}\bar{\gamma}_n U_{n,\beta}$ and the linear combination $\bar{S}_n = \sum_\alpha U_{n,\alpha} S_\alpha$.

The Lindblad equations (4.38), (4.40) and (4.41) can be formally written in the following form

$$\begin{aligned} \frac{d\hat{\rho}}{dt} &= \hat{\hat{\mathcal{L}}}\hat{\rho} \\ &= \hat{\hat{\mathcal{L}}}_H \hat{\rho} + \hat{\hat{\mathcal{L}}}_D \hat{\rho}. \end{aligned} \tag{4.42}$$

Here, we have introduced a notation with two hats to indicate a superoperator ($\hat{\hat{\mathcal{L}}}$), i.e. a linear transformation in the space of operators (here density matrices). As for operators, we will use the "hat" notation only when necessary in order to avoid confusion. In (4.42) $\hat{\hat{\mathcal{L}}}_H$ describes the Liouville von Neumann contribution and thus unitary time evolution, while $\hat{\hat{\mathcal{L}}}_D$ is the so-called dissipator. It is straightforward to prove (see, e.g. [83, 84]) that the Lindblad equation preserves the properties of the density matrix, namely

$$\mathrm{Tr}\rho = 1 \qquad \rho = \rho^\dagger \qquad \rho \geq 0\,. \tag{4.43}$$

4.4.2 Solution of the Lindblad Equation by Exact Diagonalization

The formal solution of the linear equation (4.42), for the case of a time-independent $\hat{\hat{\mathcal{L}}}$, is obtained in the usual way:

$$\hat{\rho}(t) = e^{\hat{\hat{\mathcal{L}}}t}\hat{\rho}(0). \tag{4.44}$$

Or, in terms of the eigenoperators $\hat{\rho}^{(\alpha)}$ and corresponding eigenvalues $\mathcal{L}_\alpha$ of $\hat{\hat{\mathcal{L}}}$, satisfying

$$\hat{\hat{\mathcal{L}}}\hat{\rho}^{(\alpha)} = \mathcal{L}_\alpha\hat{\rho}^{(\alpha)}, \tag{4.45}$$

one has

$$\hat{\rho}(t) = \sum_\alpha c_\alpha\hat{\rho}^{(\alpha)}e^{\mathcal{L}_\alpha t}, \tag{4.46}$$

where the c_α are fixed by the initial $t = 0$ condition. Since $\hat{\hat{\mathcal{L}}}$ is *non-Hermitian* its eigenvalues are complex, so we write them as

$$\mathcal{L}_\alpha = R_\alpha + iI_\alpha. \tag{4.47}$$

From (4.46) we readily see that we must have $R_\alpha \leq 0$ since otherwise there would be unphysical exponential divergences at large times. The coefficients $-R_\alpha$ are the decay rates of the exponentially damped modes described by the corresponding $\hat{\rho}^{(\alpha)}$. In order for the trace to be preserved, at least one eigenvalue, say the one with $\alpha = 0$, is expected to be zero,[10] $\mathcal{L}_{\alpha=0} = 0$. Then $\hat{\rho}^{(\alpha=0)}$ corresponds to the stationary or steady state which survives in the long-time limit.

Alternatively, instead of addressing the full "doubled" many-body space of the density matrix, one can use quantum trajectory methods [82, 89–92], whereby the density matrix is replaced by an ensemble of quantum states and the dissipative terms of the Lindblad equation produces so-called quantum jumps.

[10] Or one or more must have $R_\alpha = 0$.

4.4.3 Fermionic Model Described by the Lindblad Equation

We are here interested in the situation of a fermionic central system connected to a reservoir of non-interacting fermions. We will later show that under some conditions, the action of the reservoir can be described by a correction to the system's Hamiltonian (Lamb shift) plus a dissipator [cf. (4.41)]

$$\begin{aligned}\mathcal{L}_D\rho =& \sum_{i,j} 2\Gamma_{1ij}\left(a_j\rho a_i^\dagger - \frac{1}{2}\left\{a_i^\dagger a_j, \rho\right\}\right)\\ &+\sum_{i,j} 2\Gamma_{2ij}\left(a_i^\dagger \rho a_j - \frac{1}{2}\left\{a_j a_i^\dagger, \rho\right\}\right).\end{aligned} \tag{4.48}$$

Here, ρ is the reduced density matrix of the central system. This becomes an exact description of the reservoir in particular limits, as discussed below. The terms in (4.48) proportional to Γ_{1ij} with jump operators a_j describe particles jumping from the central system into the reservoir. The ones with Γ_{2ij} describe the opposite process, namely particles jumping from the reservoir into the central system.

Example As an example, consider a single-level model [(4.48) with no indices i and j], for which the Hamiltonian reads

$$H = \varepsilon a^\dagger a. \tag{4.49}$$

By explicitly solving for the steady-state of the Lindblad equation it is straightforward to show that the steady state occupation reads

$$\langle a^\dagger a\rangle = \frac{\Gamma_2}{\Gamma_1 + \Gamma_2} \tag{4.50}$$

4.4.4 Microscopic Derivation of the Lindblad Equation

In this section we will provide an explicit derivation of (4.41) starting from a microscopic model describing a central system coupled to a reservoir. In (4.41), ρ is the reduced density matrix of the central system after tracing out the reservoir. This topic has been treated in a number of textbooks. Here, we roughly follow [84, 88].

We start from a "universe" consisting of a central system +reservoir and described by the following Hamiltonian

$$H = H_S \otimes I + I \otimes H_R + V = H_0 + V. \tag{4.51}$$

Here, H_S (H_R) is the Hamiltonian for the isolated central system (reservoir), and V is the coupling between the two. The latter can always be expressed in terms of a sum of tensor products of system (S_α) and reservoir (R_α) operators:

$$V = v \sum_\alpha S_\alpha \otimes R_\alpha . \tag{4.52}$$

The parameter v is introduced for convenience as a measure for the strength of the coupling V, and will be used later in order to discuss the range of validity of the Lindblad equation. v is chosen in such a way that the operators S_α and R_α are of order 1. The full density matrix ρ for the (closed) universe obeys the Liouville-von Neumann equation

$$\dot{\rho} = -i\,[H_0 + V, \rho] . \tag{4.53}$$

The goal is to integrate out the reservoir degrees of freedom in order to arrive at an effective time evolution equation for the reduced density matrix

$$\rho_S(t) = \mathrm{Tr}_R \rho(t), \tag{4.54}$$

which only depends on system operators and $\rho_S(t)$ itself. As we will show on the next pages, under certain conditions one gets an equation of Lindblad form

$$\frac{d\hat{\rho}_S(t)}{dt} = \hat{\hat{\mathcal{L}}}\hat{\rho}_S(t). \tag{4.55}$$

A central aspect is that this equation is time local, which is a consequence of the so-called *Markovian* assumption for the reservoirs' dynamics, see below, so that memory effects are neglected.

A trivial limit is the decoupled case $V = 0$. Here, the time evolution for $\rho_S(t)$ is unitary and the Lindblad equation is given by

$$\frac{d}{dt}\hat{\rho}_S = -i\left[\hat{H}_S, \hat{\rho}_S\right] . \tag{4.56}$$

Introducing the density matrix in the interaction picture

$$\bar{\rho}(t) \equiv e^{iH_0 t}\rho(t)e^{-iH_0 t}, \tag{4.57}$$

Equation (4.53) can be rewritten as

$$\dot{\bar{\rho}} = -i\left[\bar{V}(t), \bar{\rho}\right], \tag{4.58}$$

where

$$\begin{aligned}\bar{V}(t) &= e^{iH_0 t} V e^{-iH_0 t} \\ &= v \sum_\alpha S_\alpha(t) \otimes R_\alpha(t),\end{aligned} \tag{4.59}$$

is the system-reservoir coupling in the interaction picture, and the time evolution of $S_\alpha(t)$[11] is determined by its corresponding unperturbed Hamiltonian, i.e. $S_\alpha(t) = e^{iH_St}S_\alpha e^{-iH_St}$, and similarly for R_α.

4.4.4.1 Born Markov Approximation

According to (4.58), the time evolution for a small step Δt is given by

$$\bar{\rho}(t+\Delta t) = -i\int_t^{t+\Delta t}\big[\bar{V}(t'), \bar{\rho}(t')\big]dt' + \bar{\rho}(t). \tag{4.60}$$

This equation can be iterated by inserting for $\bar{\rho}(t')$ again (4.60), leading to

$$\Delta\bar{\rho}(t) = -i\int_t^{t+\Delta t}\big[\bar{V}(t'), \bar{\rho}(t)\big]dt' - \int_t^{t+\Delta t}dt'\int_t^{t'}dt''\Big[\bar{V}(t'), \Big[\bar{V}(t''), \bar{\rho}(t'')\Big]\Big], \tag{4.61}$$

where $\Delta\bar{\rho}(t) \equiv \bar{\rho}(t+\Delta t) - \bar{\rho}(t)$. We now split $\bar{\rho}$ in the following way:

$$\bar{\rho}(t) = \bar{\rho}_S(t)\otimes\bar{\rho}_{R0}(t) + \delta\bar{\rho}_{corr}(t), \tag{4.62}$$

where, $\bar{\rho}_S(t) \equiv \mathrm{Tr}_R\bar{\rho}(t)$ is the system's reduced density matrix, $\bar{\rho}_{R0}(t)$ the unperturbed ($V = 0$) reservoir density matrix, and $\delta\bar{\rho}_{corr}$ the rest, which accounts for correlations between system and reservoir. $\bar{\rho}_{R0}(t)$ is chosen to commute with H_R, so that it is time independent $\bar{\rho}_{R0}(t) \to \rho_R$. This is not a major restriction and, for example, this is the case for the equilibrium distribution $\rho_R \propto e^{-\beta H_R}$.

The main approximation now will be to neglect $\delta\bar{\rho}_{corr}$. In Sect. 4.4.4.3 we will discuss under which conditions and in what sense this is justified. With this approximation, (4.61) becomes

$$\begin{aligned}\Delta\bar{\rho}_S(t) = &-i\int_t^{t+\Delta t}dt'\mathrm{Tr}_R\left[\bar{V}(t'), \bar{\rho}_S(t')\otimes\rho_R\right] \\ &- \int_t^{t+\Delta t}dt'\int_t^{t'}dt''\mathrm{Tr}_R\Big[\bar{V}(t'), \Big[\bar{V}(t''), \bar{\rho}_S(t'')\otimes\rho_R\Big]\Big].\end{aligned} \tag{4.63}$$

The first term on the r.h.s. of (4.63) can generally be taken to be zero. Specifically, this part contains terms of the form

$$\mathrm{Tr}_R R_\alpha(t')\rho_R = \mathrm{Tr}_R R_\alpha\rho_R = r_\alpha, \tag{4.64}$$

[11]We omit the bar used to indicate the interaction picture here, since this is already clear from the time dependence.

and the numbers r_α can be chosen without restriction to be zero. Indeed, for nonzero r_α one may introduce new reservoir operators

$$R'_\alpha = R_\alpha - r_\alpha \mathbb{1}, \tag{4.65}$$

which yield $\text{Tr}_R R'_\alpha \rho_R = 0$. The coupling Hamiltonian (4.52) becomes

$$V = v \sum_\alpha \left(R'_\alpha \otimes S_\alpha + r_\alpha S_\alpha \right), \tag{4.66}$$

and the term $v \sum_\alpha r_\alpha S_\alpha$ can be reabsorbed into H_S.

To get a useful expression out of the remaining term in (4.63) (second line), we need to introduce the *Markov* approximation. In order to understand it, let us denote by T_S the time scale over which the system, i.e. $\bar{\rho}_S$, changes due to the interaction with the environment. In terms of $\bar{\rho}_S$ this is clearly given by

$$T_S \sim \left(\frac{|\Delta \bar{\rho}_S|}{\Delta t} \frac{1}{|\bar{\rho}_S|} \right)^{-1}, \tag{4.67}$$

where $| \cdots |$ is some suitable measure. We now take Δt, which up to now can be chosen arbitrarily, to be

$$\Delta t \ll T_S. \tag{4.68}$$

In this way, since the variation of $\bar{\rho}_S(t'')$ for $t \leq t'' \leq t + \Delta t$ is negligible, one can replace in (4.63) $\bar{\rho}_S(t'') \to \bar{\rho}_S(t)$. With this one obtains from (4.63) the coarse-grained derivative

$$\frac{\Delta \bar{\rho}_S(t)}{\Delta t} = -\frac{1}{\Delta t} \int_t^{t+\Delta t} dt' \int_t^{t'} dt'' \, \text{Tr}_R \left[\bar{V}(t'), \left[\bar{V}(t''), \bar{\rho}_S(t) \otimes \rho_R \right] \right]. \tag{4.69}$$

This equation looks *Markovian*, as $\bar{\rho}_S(t)$ is time local and there is no memory on the past. However, this is valid only in a very small interval Δt, and we will see below that Δt cannot be taken arbitrarily small.

4.4.4.2 Reservoir Correlation Functions

Equation (4.69) contains terms of the form

$$\text{Tr}_R R_\alpha(t') R_\beta(t'') \rho_R \equiv C_{\alpha\beta}(t' - t''), \tag{4.70}$$

and permutations thereof. Here, we have again exploited time translational invariance of the reservoir. Provided the reservoir is infinite, its correlation functions $C_{\alpha\beta}(\tau)$

decay with a characteristic time scale τ_R. As we will see, in order to be able to neglect $\delta\bar{\rho}_{corr}$, one must have

$$\tau_R \ll \Delta t. \tag{4.71}$$

This has to be supplemented with the previous requirement $\Delta t \ll T_S$. More specifically, we will see (cf. [88]) that if (4.71) is fulfilled, the contribution from $\delta\bar{\rho}_{corr}$ are canceled from coarse graining on the scale Δt.

Introducing the time difference $\tau = t' - t'' \in (0, \Delta t)$, the integrals in (4.69) can be rewritten as

$$\int_0^{\Delta t} d\tau \int_{t+\tau}^{t+\Delta t} dt' \ldots. \tag{4.72}$$

The integrand contains terms $C_{\alpha\beta}(\tau)$, which decay in a time $\tau_R \ll \Delta t$. Therefore, it is safe to change the boundaries of the integrals to

$$\int_0^{\infty} d\tau \int_{t}^{t+\Delta t} dt' \ldots \tag{4.73}$$

With the explicit form of the coupling Hamiltonian (4.52), (4.69) becomes

$$\begin{aligned}\frac{\Delta\bar{\rho}_S(t)}{\Delta t} = -\int_0^{\infty} d\tau \frac{1}{\Delta t}\int_t^{t+\Delta t} dt' v^2 \sum_{\alpha\beta} \Big\{ & S_\alpha(t')S_\beta(t'-\tau)\bar{\rho}_S(t)C_{\alpha\beta}(\tau) \\ & - S_\alpha(t')\bar{\rho}_S(t)S_\beta(t'-\tau)C_{\beta\alpha}(-\tau) \\ & - S_\beta(t'-\tau)\bar{\rho}_S(t)S_\alpha(t')C_{\alpha\beta}(\tau) \\ & + \bar{\rho}_S(t)S_\beta(t'-\tau)S_\alpha(t')C_{\beta\alpha}(-\tau)\Big\}.\end{aligned} \tag{4.74}$$

From here we shall omit to explicitly indicate the time argument for $\bar{\rho}_S(t)$. Equations (4.74) in turn can be rewritten in terms of commutators as

$$\begin{aligned}\frac{\Delta\bar{\rho}_S}{\Delta t} = -\int_0^{\infty} d\tau \frac{1}{\Delta t}\int_t^{t+\Delta t} dt' v^2 \sum_{\alpha\beta} \Big\{ & C_{\alpha\beta}(\tau)\Big[S_\alpha(t'), S_\beta(t'-\tau)\bar{\rho}_S\Big] \\ & + C_{\beta\alpha}(-\tau)\Big[\bar{\rho}_S S_\beta(t'-\tau), S_\alpha(t')\Big]\Big\},\end{aligned} \tag{4.75}$$

and we are now in the position to determine the order of magnitude of (4.75). As stated above, τ_R is assumed to be the characteristic decay time of $C_{\alpha\beta}(\tau)$, (4.70). Since the involved operators R_α and ρ_R are of $O(1)$, one can estimate

$$\int_0^{\infty} d\tau\, C_{\alpha\beta}(\tau) \sim \tau_R. \tag{4.76}$$

Since also the $S_\alpha \sim O(1)$ we can estimate

$$\frac{1}{T_S} \sim \frac{|\Delta\bar{\rho}_S|}{\Delta t}\frac{1}{|\bar{\rho}_S|} \sim \tau_R v^2. \tag{4.77}$$

The two conditions (4.68) and (4.71) become

$$\tau_R \ll \Delta t \ll \frac{1}{\tau_R v^2}, \tag{4.78}$$

which brings us to the necessary condition

$$\tau_R\, v \ll 1. \tag{4.79}$$

In terms of energy scales (4.78) reads

$$W_R \equiv \frac{1}{\tau_R} \gg \frac{1}{\Delta t} \gg v^2\tau_R \equiv \Gamma_S = \frac{1}{T_S}. \tag{4.80}$$

Here, W_R is the typical energy scale of the reservoir controlling its relaxation rate, e.g. the bandwidth or chemical potential μ, and Γ_S is a measure for the system-reservoir coupling which will be related to the system's relaxation rate. From (4.79) we have the requirement

$$W_R \gg v. \tag{4.81}$$

A further scale is the typical spacing $\Delta\varepsilon_S$ of the system's energies. Depending on its magnitude there can be two situations

(1) $\Delta\varepsilon_S \gg \Gamma_S \to$ *weak coupling* limit: One then takes

$$\Delta\varepsilon_S \gg \frac{1}{\Delta t} \gg \Gamma_S\,, \tag{4.82}$$

which leads to the so-called secular approximation, [83] which we are not going to discuss here.

(2) $W_R \gg \Delta\varepsilon_S \to$ *singular coupling* limit: Formally this is obtained by rewriting (4.51) as

$$H = H_S + \frac{1}{\delta}V + \frac{1}{\delta^2}H_R, \tag{4.83}$$

and taking $\delta \to 0$.

Of course one can, in principle, have both situations at the same time, provided

$$W_R \gg \Delta\varepsilon_S \gg \frac{1}{\Delta t} \gg \Gamma_S. \tag{4.84}$$

In these lectures we focus on the second case. The interesting situation is especially when $\Gamma_S \sim \Delta\varepsilon_S$, so that the action of the environment on the system cannot be regarded as small. Here we have

$$\frac{1}{\tau_R} \equiv W_R \gg \frac{1}{\Delta t} \gg \Delta\varepsilon_S \sim \Gamma_S. \tag{4.85}$$

We now return to the evolution equation for $\bar{\rho}_S(t)$, (4.75). Let us consider the eigenvectors $|n\rangle$ with eigenvalues ε_n of the system Hamiltonian in terms of which the time dependence of the system operators can be rewritten as

$$S_\alpha(t') = \sum_{m,n} |n\rangle\langle m| \langle n|S_\alpha|m\rangle e^{i(\varepsilon_n-\varepsilon_m)t'}, \tag{4.86}$$

see also (4.59). The integration in (4.75) has to be evaluated in the range $\tau \in (0, \tau_R)$ and $t' - t \in (0, \Delta t)$, which allows one to approximate

$$(\varepsilon_n - \varepsilon_m)t' \sim (\varepsilon_n - \varepsilon_m)(t' - \tau) \sim (\varepsilon_n - \varepsilon_m)t, \tag{4.87}$$

since $\Delta\varepsilon_S\ \Delta t \ll 1$ due to (4.85). Therefore, the detailed t'- and τ-dependence of S_α can be neglected and we can replace $S_\alpha(t')$ and $S_\alpha(t' - \tau)$ in (4.75) by $S_\alpha(t)$. This allows us to pull out the t-dependent terms and the integration $\frac{1}{\Delta t}\int_t^{t+\Delta t} dt' \to 1$ can be dropped. We denote the remaining integrals over the reservoir correlation functions by

$$C^{\pm}_{\alpha\beta} \equiv \int_0^{\infty} C_{\alpha\beta}(\pm\tau)\mathrm{d}\tau. \tag{4.88}$$

Furthermore, one can formally interpret $\Delta\bar{\rho}_S/\Delta t$ on the lhs of (4.75) as a derivative $\mathrm{d}\bar{\rho}_S/\mathrm{d}t$. The t-dependent terms in (4.75) are of the form

$$\begin{aligned} S_\alpha(t)S_\beta(t)\bar{\rho}_S(t) &= \underbrace{e^{iH_St}S_\alpha e^{-iH_St}}_{S_\alpha(t)}\ e^{iH_St}S_\beta\ \underbrace{e^{-iH_St}\bar{\rho}_S(t)e^{iH_St}}_{\rho_S(t)}\ e^{-iH_St} \\ &= e^{iH_St}S_\alpha S_\beta\rho_S(t)e^{-iH_St}. \end{aligned} \tag{4.89}$$

We now transform the derivative from the interaction to the Schrödinger representation. From differentiating

$$\bar{\rho}_S(t) = e^{iH_St}\rho_S(t)e^{-iH_St}, \tag{4.90}$$

one finds

$$\frac{d\bar{\rho}_S(t)}{dt} = e^{iH_S t}\left(i\,[H_S,\rho_S] + \frac{d\rho_S}{dt}\right)e^{-iH_S t}, \tag{4.91}$$

$$\Rightarrow \frac{d\rho_S}{dt} = -i\,[H_S,\rho_S] + e^{-iH_S t}\frac{d\bar{\rho}_S(t)}{dt}e^{iH_S t}, \tag{4.92}$$

where we omitted the time argument of $\rho_S(t)$, and the terms $e^{-iH_S t}\ldots e^{iH_S t}$ cancel the ones $e^{iH_S t}\ldots e^{-iH_S t}$ in (4.89). We thus get from (4.75) and (4.92)

$$\frac{d\rho_S}{dt} = -i\,[H_S,\rho_S] - v^2\sum_{\alpha\beta}\left(C^+_{\alpha\beta}\left[S_\alpha, S_\beta\rho_S\right] + C^-_{\alpha\beta}\left[\rho_S S_\alpha, S_\beta\right]\right), \tag{4.93}$$

with $C^\pm_{\alpha\beta}$ given in (4.88). Furthermore, by defining

$$C_{\alpha\beta} \equiv C^+_{\alpha\beta} + C^-_{\alpha\beta} = \int_{-\infty}^{\infty} C_{\alpha\beta}(\tau)d\tau \tag{4.94}$$

$$\bar{C}_{\alpha\beta} \equiv C^+_{\alpha\beta} - C^-_{\alpha\beta} = \int_{-\infty}^{\infty} \mathrm{sgn}(\tau)C_{\alpha\beta}(\tau)d\tau, \tag{4.95}$$

one arrives at

$$\begin{aligned}\frac{d\rho_S}{dt} =& -i\,[H_S,\rho_S] + v^2\sum_{\alpha,\beta} C_{\alpha\beta}\left(S_\beta\rho_S S_\alpha - \frac{1}{2}\{S_\alpha S_\beta,\rho_S\}\right)\\ &- v^2\sum_{\alpha,\beta}\bar{C}_{\alpha\beta}\frac{1}{2}\left[S_\alpha S_\beta,\rho_S\right].\end{aligned} \tag{4.96}$$

This expression can be rewritten in a more convenient form when explicitly considering that the coupling Hamiltonian V in (4.52) is Hermitian and, thus, can be rewritten as

$$V = v\sum_\alpha S_\alpha\otimes R_\alpha \tag{4.97}$$

$$= v{\sum_\alpha}' \left(S_\alpha\otimes R_\alpha + S^\dagger_\alpha\otimes R^\dagger_\alpha\right), \tag{4.98}$$

where the ${\sum}'_\alpha$ is such that the two expressions coincide. Introducing $\overline{\alpha}$ indices in such a way that

$$S_{\overline{\alpha}} = S^\dagger_\alpha \qquad R_{\overline{\alpha}} = R^\dagger_\alpha, \tag{4.99}$$

as well as new coefficients

$$\gamma_{\alpha\beta} \equiv C_{\overline{\alpha}\beta} v^2, \qquad i\sigma_{\alpha\beta} \equiv -v^2 \bar{C}_{\alpha\overline{\beta}}, \tag{4.100}$$

we can rewrite (4.96) as (we omit the prime from the sum from now on)

$$\begin{aligned} \frac{\mathrm{d}\rho_S}{\mathrm{d}t} &= -i\left[\widetilde{H}_S, \rho_S\right] + \sum_{\alpha\beta} \gamma_{\alpha\beta}\left(S_\beta \rho_S S_\alpha^\dagger - \frac{1}{2}\{S_\alpha^\dagger S_\beta, \rho_S\}\right) \\ &\equiv \mathcal{L}_H \rho_S + \mathcal{L}_D \rho_S. \end{aligned} \tag{4.101}$$

Here,

$$\widetilde{H}_S = H_S + \frac{1}{2}\sum_{\alpha\beta} \sigma_{\alpha\beta} S_\alpha S_\beta^\dagger. \tag{4.102}$$

Equation (4.101) is just the *Lindblad* equation (4.41) stated before, provided the coefficient matrix $\gamma_{\alpha\beta}$ is Hermitian and semipositive definite and $\sigma_{\alpha\beta}$ is Hermitian, which is straightforward to prove. As a side remark, the same form of the Lindblad equation is obtained in the weak-coupling limit (4.82), see, e.g. [83, 84, 87].

As mentioned before, $\mathcal{L} = \mathcal{L}_H + \mathcal{L}_D$ is the Lindblad superoperator. It consists of a unitary part $\mathcal{L}_H$, which simply provides a correction [the so-called "Lamb shift" cf. (4.102)] to the system Hamiltonian, and of the dissipator $\mathcal{L}_D$. Furthermore, notice that the Lindblad equation is Markovian since $\mathrm{d}\rho_S(t)/\mathrm{d}t$ only depends on $\rho_S(t)$, i.e. there are no contribution from the past values of $\rho_S(t)$.

4.4.4.3 Validity of Neglecting $\delta\bar{\rho}_{\mathrm{corr}}$

In order to derive the pleasant equation (4.101) we introduced the quite drastic approximation of neglecting correlations between system and environment described by $\delta\bar{\rho}_{\mathrm{corr}}$. Fortunately, one can readily show that this is justified without the need to introduce further assumptions beyond the ones we have already made in (4.85).

We follow the discussion of [88]. The correction term, as defined in (4.62), accounts for both correlations as well as changes in ρ_R. When including $\delta\bar{\rho}_{\mathrm{corr}}$ in (4.63), it enters in the first term on the r.h.s. and leads to the modification $\delta\Delta\bar{\rho}_S$, of $\Delta\bar{\rho}_S$:

$$\delta\Delta\bar{\rho}_S = -i\,\mathrm{Tr}_R \int_t^{t+\Delta t} \left[\bar{V}(t'), \delta\bar{\rho}_{\mathrm{corr}}(t)\right] \mathrm{d}t'. \tag{4.103}$$

Let us consider some initial time $t_0 < t$ at which $\delta\bar{\rho}_{\mathrm{corr}}(t_0) = 0$, e.g. $t_0 \to -\infty$. During the time evolution this term becomes nonzero due to V, and to first order in v we have

$$\delta \bar{\rho}_{\text{corr}}(t) \sim \int_{-\infty}^{t} dt'' \bar{V}(t'')... \tag{4.104}$$

Upon insertion into (4.103) one finds terms of the form

$$\delta \Delta \bar{\rho}_S \sim \int_t^{t+\Delta t} dt' \int_{-\infty}^{t} \underbrace{\text{Tr}_R \left(\bar{V}(t') \bar{V}(t'') \rho_R \right)}_{v^2 \langle R(t') R(t'') \rangle S(t') S(t'')} dt''... \tag{4.105}$$

With (4.52) and (4.70) we can relate this to the reservoir correlation functions

$$v^2 \langle R(t') R(t'') \rangle \sim C_{\alpha\beta}(t' - t'') v^2, \tag{4.106}$$

which are nonzero only in a small region $|t' - t''| < \tau_R$. This allows one to estimate

$$\delta \Delta \bar{\rho}_S \sim \int_t^{t+\tau_R} dt' \int_{t-\tau_R}^{t} dt'' C_{\alpha\beta}(t' - t'') v^2 \tag{4.107}$$

$$\sim v^2 \tau_R^2, \tag{4.108}$$

which has to be compared with

$$\Delta \bar{\rho}_S \sim \frac{\Delta t}{T_S} \sim \Delta t \, \tau_R \, v^2. \tag{4.109}$$

Therefore, the condition for neglecting the contribution $\delta \Delta \bar{\rho}_S$ originating from δ becomes

$$\frac{\delta \Delta \bar{\rho}_S}{\Delta \bar{\rho}_S} \sim \frac{\tau_R}{\Delta t} \ll 1, \tag{4.110}$$

i.e. (4.71). In other words, averaging over a time $\Delta t \gg \tau_R$ allows one to "forget" the effects of correlations prior to t.

4.4.5 Derivation for a Fermionic System-Reservoir Setup

We now derive the Lindblad equation (4.48) from a microscopic fermion-reservoir model and discuss the limit in which the Lindblad representation of the reservoir becomes exact. According to (4.79) we need $W_R = \frac{1}{\tau_R} \gg v$, which is fulfilled when

(1) The DOS of the reservoir is ω-independent, i.e., the so-called *wide-band limit*

and

(2) T and/or $|\mu| \to \infty$, which corresponds to an ω-independent reservoir occupation.

We consider a generic noninteracting fermionic reservoir described by the Hamiltonian

$$
\begin{aligned}
H_R &= \sum_k \varepsilon_k c_k^\dagger c_k, \\
V &= \sum_{kn} v_{kn} \left(c_k^\dagger a_n + h.c. \right),
\end{aligned} \tag{4.111}
$$

where $c_k^{(\dagger)}$ are reservoir and $a_n^{(\dagger)}$ system fermionic operators, and v_{kn} are real-valued coupling constants. We don't need to specify the form of the system Hamiltonian $H_S[a]$, since we just want to derive the effects of the reservoir. The reservoir levels ε_k must be continuous in order to produce dissipation, so we will let the level spacing go to zero, $\Delta\varepsilon \to 0$, and we introduce continuous operators[12]

$$
c_k = \sqrt{\frac{\Delta\varepsilon}{2\pi}} c(\varepsilon)\,, \tag{4.112}
$$

and couplings

$$
v_{kn} = \sqrt{\frac{\Delta\varepsilon}{2\pi}} v_n(\varepsilon)\,, \tag{4.113}
$$

where the system indices n remain discrete. In this way, the reservoir and coupling Hamiltonians become

$$
\begin{aligned}
H_R &= \int \frac{d\varepsilon}{2\pi}\, \varepsilon\, c^\dagger(\varepsilon) c(\varepsilon), \\
V &= \sum_n \int \frac{d\varepsilon}{2\pi}\, v_n(\varepsilon)\, c^\dagger(\varepsilon) a_n + h.c. \\
&\equiv \sum_n R_n a_n + h.c.
\end{aligned} \tag{4.114}
$$

This is in the form of (4.52), except for the fact that, for simplicity, we have absorbed v in the definition of the R_n. From (4.114) we read off

$$
R_n = \int \frac{d\varepsilon}{2\pi} v_n(\varepsilon) c^\dagger(\varepsilon), \tag{4.115}
$$

We need to evaluate the reservoir correlation functions (4.70)

$$
c_{\bar{n}m}(\tau) = \langle R_n^\dagger(\tau) R_m(0) \rangle, \tag{4.116}
$$

[12]To be more rigorous: $c(\varepsilon) = \frac{1}{\sqrt{D(\varepsilon)}} \sum_k \delta(\varepsilon - \varepsilon_k) c_k$.

and

$$\begin{aligned} c_{n\bar{m}}(\tau) &= \langle R_n(\tau) R_m^\dagger(0) \rangle \\ &= \int \frac{d\varepsilon\, d\varepsilon'}{(2\pi)^2}\, v_n(\varepsilon) v_m(\varepsilon') \langle c^\dagger(\varepsilon, \tau) c(\varepsilon') \rangle, \end{aligned} \tag{4.117}$$

where

$$c^\dagger(\varepsilon, \tau) = e^{-i\varepsilon\tau} c^\dagger(\varepsilon) . \tag{4.118}$$

The occupation of reservoir states is given by

$$\begin{aligned} \langle c_k^\dagger c_{k'} \rangle &= \delta_{kk'}\, n(k) \\ &= \frac{\Delta\varepsilon}{2\pi} \langle c^\dagger(\varepsilon) c(\varepsilon') \rangle, \end{aligned} \tag{4.119}$$

and with $\delta_{kk'}/\Delta\varepsilon \to \delta(\varepsilon - \varepsilon')$ one finds that

$$\langle c^\dagger(\varepsilon, \tau) c(\varepsilon') \rangle = 2\pi \delta(\varepsilon - \varepsilon') n(\varepsilon) e^{-i\varepsilon\tau} , \tag{4.120}$$

by which (4.117) simplifies to

$$c_{n\bar{m}}(\tau) = \int \frac{d\varepsilon}{2\pi}\, v_n(\varepsilon) v_m(\varepsilon)\, n(\varepsilon)\, e^{-i\varepsilon\tau} . \tag{4.121}$$

As discussed in (4.80), in order for the Lindblad equation representation of the reservoir to be accurate, the correlation functions (4.121) must decay fast enough, i.e. with a rate $1/\tau_R$ much larger than the $\Delta\varepsilon_S$ and v. $1/\tau_R$ is proportional to the width of the argument in (4.121), $F(\varepsilon) \equiv v_n(\varepsilon) v_m(\varepsilon) n(\varepsilon)$. Therefore, strictly speaking the Lindblad representation becomes exact when $F(\varepsilon)$ is constant. In this case,

$$c_{n\bar{m}}(\tau) \propto \delta(\tau) , \tag{4.122}$$

i.e. the Markovian condition. It is interesting to notice that this is the only requirement and once (4.122) is fulfilled there is no further weak-coupling requirement although a weak-coupling expansion was used for the Born-Markov approximation. The condition $F(\varepsilon) = \text{const.}$ requires both the wide-band limit $v_n(\varepsilon) = \text{const.}$ and $n(\varepsilon) = \text{const.}$. The latter corresponds to having either (i) $\mu \to \pm\infty$ or (ii) $T \to \infty$. Otherwise, $c_{n\bar{m}}(\tau)$ decays with a rate $1/\tau_R$ proportional to the width of $F(\varepsilon)$. In nonequilibrium situations it is useful to have reservoirs with different occupations $n(\varepsilon)$. This is not in contradiction with the above condition since one can generalize (4.114) by including a sum over separate reservoirs α with constant but different occupations $n_\alpha(\varepsilon)$.

From (4.121) we determine the correlation functions (4.95), (4.100) by exploiting the following relations

$$I_1 \equiv \int_0^{\infty} e^{-i(\varepsilon - i\delta)\tau} \mathrm{d}\tau = \frac{1}{i(\varepsilon - i\delta)} = -i\mathrm{P}\frac{1}{\varepsilon} + \pi\delta(\varepsilon),$$
$$I_2 \equiv \int_{-\infty}^{0} e^{-i(\varepsilon + i\delta)\tau} \mathrm{d}\tau = -\frac{1}{i(\varepsilon + i\delta)} = I_1^* . \tag{4.123}$$

This gives for the two matrices γ and σ of (4.101) and (4.102)

$$c_{n\bar{m}} = \int \mathrm{d}\tau \; c_{n\bar{m}}(\tau) = v_n(0) v_m(0) n(0) = \gamma_{\bar{n}\bar{m}},$$
$$\bar{c}_{n\bar{m}} = \int \mathrm{d}\tau \; c_{n\bar{m}}(\tau) \,\mathrm{sgn}(\tau) = -i\mathrm{P} \int \frac{\mathrm{d}\varepsilon}{\pi} \frac{1}{\varepsilon} v_n(\varepsilon) v_m(\varepsilon) n(\varepsilon) = -i\sigma_{nm} . \tag{4.124}$$

Even for energy independent $v_n(\varepsilon)v_m(\varepsilon)n(\varepsilon)$, the quantities σ_{nm} may be sensitive to their values at high energies. For simplicity, we here take even functions $v(\varepsilon)$, $n(\varepsilon)$, so that $\sigma_{nm} = 0$. Similarly

$$\gamma_{nm} = v_n(0) v_m(0) m(0) \qquad (m(\varepsilon) = 1 - n(\varepsilon)),$$
$$\sigma_{\bar{n}\bar{m}} = -\mathrm{P} \int \frac{\mathrm{d}\varepsilon}{\pi} \frac{1}{\varepsilon} v_n(\varepsilon) v_m(\varepsilon) m(\varepsilon) . \tag{4.125}$$

Thus, the parameters entering (4.48) are (we omit the ε-dependence of the v_n and of n)

$$2\Gamma_{1nm} = \gamma_{nm} = v_n v_m (1 - n),$$
$$2\Gamma_{2nm} = \gamma_{\bar{n}\bar{m}} = v_n v_m n. \tag{4.126}$$

As already discussed, Γ_{1nm} describes the removal of particles from the system which is consistent with it being proportional to $(1 - n)$, and Γ_{2nm} describes particle injection and is proportional to n.

For the 1-level model discussed above, we have in the steady state [cf. (4.50)]

$$\langle a^{\dagger} a \rangle = \frac{\Gamma_2}{\Gamma_1 + \Gamma_2} = n, \tag{4.127}$$

which we expect for a level in equilibrium with a reservoir.

4.5 Superfermion Representation

The so-called superfermion representation is a useful scheme to map the Lindblad equation onto a standard operator problem, in which the superoperator $\hat{\hat{\mathcal{L}}}$ acting on $\hat{\rho}$ is replaced by an ordinary operator $\hat{\mathcal{L}}$ acting on the corresponding state vector $|\rho\rangle$ in an enlarged Hilbert space. Like (4.42), the resulting equation is of "Schrödinger" type

$$\frac{\mathrm{d}}{\mathrm{d}t}|\rho\rangle = \hat{\mathcal{L}}|\rho\rangle, \tag{4.128}$$

in which, however, the "Hamiltonian" $i\hat{\mathcal{L}}$ is a non-Hermitian operator.

Here, we follow the treatment by Dzhioev and Kosov [65], see also [93, 94], as well as [66, 67] for an earlier treatment. The starting point is an augmented fermion Fock space, in which the original Hilbert space is doubled. Starting from the basis states $|n\rangle$ of the original space of dimension $N_{\mathcal{H}}$, one introduces additional"tilde" states and defines the new basis states $|n\rangle|\widetilde{m}\rangle$. The size of the new Hilbert space clearly becomes $N_{\mathcal{H}} \to N_{\mathcal{H}}^2$. This allows for a convenient representation of (system) density matrices[13]:

$$\hat{\rho} = \sum_{nm} |n\rangle \underbrace{\langle m|}_{\Rightarrow|\tilde{m}\rangle} \rho_{nm}. \tag{4.129}$$

For this one introduces the so-called "left vacuum"

$$|I\rangle = \sum_{m} |m\rangle|\widetilde{m}\rangle \,, \tag{4.130}$$

which is essentially a purification of the identity operator. Applying $\hat{\rho}$ to the left vacuum maps the density matrix onto a state vector of the augmented space.

$$\hat{\rho} \Rightarrow |\rho\rangle = \hat{\rho} \otimes \underbrace{\widetilde{I}}_{\text{implicit}} |I\rangle = \sum_{nm} \rho_{nm}|n\rangle|\widetilde{m}\rangle. \tag{4.131}$$

In general, for an arbitrary operator $\hat{B}$ one defines the corresponding state vector

$$|B\rangle \equiv \hat{B}|I\rangle \,, \tag{4.132}$$

which can be used to evaluate traces of operators

$$\mathrm{Tr}\hat{B} = \langle I|\hat{B}|I\rangle = \langle I|B\rangle. \tag{4.133}$$

[13] We omit here the system index of ρ_S for the sake of clarity.

Proof

$$\langle I|\hat{B}|I\rangle = \sum_{knm} \langle k|\langle \widetilde{k}|B_{nm}|n\rangle|\widetilde{m}\rangle = \sum_{k} B_{kk}. \tag{4.134}$$

In particular, expectation values of operators are given by

$$\begin{aligned}\langle \hat{A}\rangle = \mathrm{Tr}\hat{A}\hat{\rho} &= \langle I| \underbrace{\hat{A}\hat{\rho}}_{(\text{implicitly } \hat{A}\hat{\rho}\otimes\widetilde{I})} |I\rangle \\ &= \langle I|\hat{A}|\rho\rangle. \end{aligned}\tag{4.135}$$

Besides expressions of the form $\hat{A}\hat{\rho}$ we also need to evaluate $\hat{\rho}\hat{A}$, which occurs for instance in the Lindblad equation. For the first case we already found that $\hat{A}\hat{\rho} \rightarrow \hat{A}|\rho\rangle$ but a transformation of the form $\hat{\rho}\hat{A} \rightarrow \hat{\rho}\hat{A}|I\rangle$ is not useful, as we would like to express also the second case in terms of an operator applied to $|\rho\rangle$. $\hat{\rho}\hat{A}$ is written as

$$\hat{\rho}\hat{A} = \sum_{nm} \rho_{nm}|n\rangle\langle m|\hat{A}, \tag{4.136}$$

i.e. $\hat{A}$ acts on the bra vector $\langle m|$. Its representation within the augmented space is thus given by

$$\hat{\rho}\hat{A} \rightarrow \sum_{nm} (\hat{\rho}\hat{A})_{nm}|n\rangle|\widetilde{m}\rangle \ . \tag{4.137}$$

One now introduces the operator[14]

$$\widetilde{A} \equiv I \otimes A^T = \sum_{kl} A_{kl}|\widetilde{l}\rangle\langle\widetilde{k}|, \tag{4.138}$$

acting on tilde states only. Applied on the state $|\rho\rangle$ it provides the desired result

$$\begin{aligned}\widetilde{A}|\rho\rangle &= \sum_{klnm} A_{kl}|\widetilde{l}\rangle\langle\widetilde{k}|\rho_{nm}|n\rangle|\widetilde{m}\rangle \\ &= \sum_{lnm} A_{ml}\rho_{nm}|n\rangle|\widetilde{l}\rangle \\ &= \sum_{nl} (\hat{\rho}\hat{A})_{nl}|n\rangle|\widetilde{l}\rangle, \end{aligned}\tag{4.139}$$

which is the r.h.s. of (4.137), i.e. we have

$$\hat{\rho}\hat{A} \rightarrow \widetilde{A}|\rho\rangle. \tag{4.140}$$

[14] Note that the definition of A^T is basis dependent.

For a Fock space of fermionic particles one has to specify the fermionic sign of each term, or in other words the ordering of the levels when specifying states such as (4.130). Considering a many-fermion system characterized by levels $i = 1, 2, \ldots, N$, which may include spin, the basis states of the two Fock spaces are indicated as

$$
\begin{aligned}
&|\underline{n}\rangle = |n_1 n_2 \ldots n_N\rangle, \\
&|\widetilde{\underline{n}}\rangle \quad \text{“tilde” Fock space,}
\end{aligned}
\tag{4.141}
$$

with corresponding creation and annihilation operators $a_i^\dagger$, a_i, $\widetilde{a}_i^\dagger$, $\widetilde{a}_i$. In the left vacuum one can include an arbitrary phase for each state. Here, it is convenient to adopt the convention

$$
\begin{aligned}
|I\rangle &= \sum_{\{\underline{n}\}} |\underline{n}, \widetilde{\underline{n}}\rangle, \\
|\underline{n}, \widetilde{\underline{n}}\rangle &= (-i)^{\sum_i n_i} (a_1^\dagger \widetilde{a}_1^\dagger)^{n_1} \ldots (a_N^\dagger \widetilde{a}_N^\dagger)^{n_N} \, |0\rangle |\widetilde{0}\rangle.
\end{aligned}
\tag{4.142}
$$

Using this expression, one obtains the so-called *tilde conjugation rules* [65]:

$$
\begin{aligned}
a_j |I\rangle &= -i \widetilde{a}_j^\dagger |I\rangle, \\
a_j^\dagger |I\rangle &= -i \widetilde{a}_j |I\rangle.
\end{aligned}
\tag{4.143}
$$

By taking their Hermitian conjugate, these can be easily generalized to

$$
\begin{aligned}
F|I\rangle &= -i \widetilde{F}^\dagger |I\rangle, \\
\langle I| F &= i \, \langle I| \widetilde{F}^\dagger,
\end{aligned}
\tag{4.144}
$$

where F is an arbitrary linear combination of a_i, $a_j^\dagger$ with real coefficients.

Proof of (4.143):

$$
\begin{aligned}
a_j |I\rangle &= \sum_{\underline{n}: n_j = 1} (-i)^{\sum_i n_i} \ldots a_j (a_j^\dagger \widetilde{a}_j^\dagger) \ldots |0\rangle |\widetilde{0}\rangle \\
&= -i \widetilde{a}_j^\dagger \sum_{\underline{n}: n_j = 0} |\underline{n}, \widetilde{\underline{n}}\rangle \quad (\text{due to } \widetilde{a}_j^\dagger,\ n_j = 0 \text{ is guaranteed}) \\
&= -i \widetilde{a}_j^\dagger |I\rangle,
\end{aligned}
$$

and similarly for $a_j^\dagger$.

4.5.1 Representation of the Lindblad Equation

For a representation of the Lindblad equation (4.101), or specifically for fermions (4.48), we have to consider the representation of different operator terms multiplying the density matrix and applied to the left vacuum state. If the S_α are linear combinations of the $a_i^{(\dagger)}$ and a_i, we have for a quadratic term multiplied from the left

$$S_\alpha S_\beta \hat{\rho}|I\rangle = S_\alpha S_\beta |\rho\rangle, \tag{4.145}$$

and for one multiplied from the right

$$\hat{\rho} S_\alpha S_\beta |I\rangle = \hat{\rho} S_\alpha \underbrace{(-i\widetilde{S}_\beta^\dagger)}_{\substack{\text{can now be moved}\\ \text{to the left}}} |I\rangle \tag{4.146}$$

$$= i\widetilde{S}_\beta^\dagger \hat{\rho}(-i\widetilde{S}_\alpha^\dagger)|I\rangle \tag{4.147}$$

$$= \widetilde{S}_\beta^\dagger \widetilde{S}_\alpha^\dagger |\rho\rangle, \tag{4.148}$$

where we used (4.143). In a similar manner, quartic terms are transformed as

$$\hat{\rho} S_1 S_2 S_3 S_4 |I\rangle = \widetilde{S}_4^\dagger \widetilde{S}_3^\dagger \widetilde{S}_2^\dagger \widetilde{S}_1^\dagger |\rho\rangle, \tag{4.149}$$

and in general, for an operator O with an even number of fermionic a_i or $a_i^\dagger$ one has

$$\hat{\rho} O |I\rangle = \widetilde{O}^\dagger |\rho\rangle. \tag{4.150}$$

Finally, (4.101) contains terms with operators multiplying on the left and on the right that become[15]

$$S_\alpha \hat{\rho} S_\beta |I\rangle = -i S_\alpha \widetilde{S}_\beta^\dagger |\rho\rangle. \tag{4.151}$$

We are now in the position to express the superfermion representation of the Lindblad equation (4.101), or more specifically of $(\hat{\hat{\mathcal{L}}}\hat{\rho})|I\rangle$. The Liouville von Neumann part $\hat{\hat{\mathcal{L}}}_H$ becomes

$$\hat{\hat{\mathcal{L}}}_H \hat{\rho}|I\rangle = -i\left[H, \hat{\rho}\right]|I\rangle = -i\left(H - \widetilde{H}\right)|\rho\rangle, \tag{4.152}$$

i.e. we have the following mapping of the superoperator $\hat{\hat{\mathcal{L}}}_H$ in the superfermion space:

$$\hat{\hat{\mathcal{L}}}_H \Rightarrow -i\left(H - \widetilde{H}\right), \tag{4.153}$$

---

[15]Note that ρ contains even products of untilded fermion operators only, and thus commutes with tilde operators.

where $\widetilde{H}$ is the Hamiltonian applied to the "tilde" part of the Hilbert space [cf. (4.138)]. Here we used (4.150) and the fact that H is Hermitian and contains terms quadratic and quartic in the fermionic operators.

The dissipator $\hat{\hat{\mathcal{L}}}_D$ in (4.101) becomes

$$\left(\hat{\hat{\mathcal{L}}}_D \hat{\rho}\right) |I\rangle = \sum_{\alpha\beta} \gamma_{\alpha\beta} \left(-i S_\beta \widetilde{S}_\alpha - \frac{1}{2} S_\alpha^\dagger S_\beta - \frac{1}{2} \widetilde{S}_\beta^\dagger \widetilde{S}_\alpha \right) |\rho\rangle, \tag{4.154}$$

where we have used (4.145, 4.148, 4.151).

On the whole, (4.152) and (4.154) transform the Lindblad equation into a "Schrödinger-type" equation governing the time evolution of the "supervector" $|\rho\rangle$,

$$\frac{\mathrm{d}}{\mathrm{d}t} |\rho\rangle = \hat{\mathcal{L}} |\rho\rangle, \tag{4.155}$$

with a non-Hermitian generator $i\hat{\mathcal{L}}$. The trace preserving property of the Lindblad equation transforms into [cf. (4.133)]

$$\frac{\mathrm{d}}{\mathrm{d}t} \mathrm{Tr} \rho = 0 \quad \Rightarrow \quad \langle I | \frac{\mathrm{d}}{\mathrm{d}t} |\rho\rangle = \langle I | \hat{\mathcal{L}} |\rho\rangle = 0. \tag{4.156}$$

Since this holds true for any $|\rho\rangle$, one has

$$\langle I | \hat{\mathcal{L}} = 0. \tag{4.157}$$

Therefore, the left vacuum $\langle I|$ is a left eigenstate of $\hat{\mathcal{L}}$ with eigenvalue zero, which explains its name. For each left eigenstate there is a right one with the same eigenvalue. In this case this is the steady state $|\rho_\infty\rangle$ with the property

$$\hat{\mathcal{L}} |\rho_\infty\rangle = 0 \, . \tag{4.158}$$

Equations of Motion

One way to address the time dependence of observables is via the *equations of motion* technique:

$$\begin{aligned} \langle A(t) \rangle &= \langle I | A | \rho(t) \rangle, \\ \frac{\mathrm{d}}{\mathrm{d}t} \langle A(t) \rangle &= \langle I | A \mathcal{L} | \rho(t) \rangle \\ &= \langle I | \underbrace{[A, \mathcal{L}]}_{\text{because } \langle I | \mathcal{L} = 0} | \rho(t) \rangle. \end{aligned} \tag{4.159}$$

In some cases, e.g. noninteracting particles, this yields a closed set of equations. In the general interacting case, however, this is not possible and a hierarchy of equations is created, which must be truncated at some point. Below we discuss in more detail an alternative way, namely to directly solve (4.155) in a manybody basis.

Example (Single-level model) Consider again a fermionic system consisting of a single level with Hamiltonian (4.49) and dissipator (4.48) with no indices i, j. Using (4.154), the superfermion representation of the Lindblad operator becomes

$$\begin{aligned}\hat{\mathcal{L}} =& - i\varepsilon\left(a^\dagger a - \widetilde{a}^\dagger\widetilde{a}\right) - \Gamma_1\left(a^\dagger a + \widetilde{a}^\dagger\widetilde{a} + 2ia\widetilde{a}\right)\\ &- \Gamma_2\left(aa^\dagger + \widetilde{a}\widetilde{a}^\dagger + 2ia^\dagger\widetilde{a}^\dagger\right),\end{aligned} \tag{4.160}$$

which can be conveniently written in a matrix form

$$\begin{aligned}\hat{\mathcal{L}} &= -i\left(a^\dagger\widetilde{a}\right)\underline{\underline{H}}\begin{pmatrix} a \\ \widetilde{a}^\dagger\end{pmatrix} + \text{const.}\\ &= -i\underline{\underline{a}}^\dagger\underline{\underline{H}}\,\underline{\underline{a}},\end{aligned} \tag{4.161}$$

with the matrix $\underline{\underline{H}}$ given by

$$\underline{\underline{H}} = \begin{pmatrix} E_+ & B \\ \bar{B} & E_-\end{pmatrix} \qquad E_\pm = \varepsilon \pm i(\Gamma_2 - \Gamma_1),\ B = 2\Gamma_2,\ \bar{B} = -2\Gamma_1\ . \tag{4.162}$$

We leave it as an exercise to use the equations of motion technique discussed above to evaluate the time dependence of the density

$$n(t) = \langle I|a^\dagger a|\rho(t)\rangle, \tag{4.163}$$

for this model.

Example (Current) Consider the single-level model (4.161), (4.162) coupled to two reservoirs, one described by the Γ_1 term and the other by the Γ_2 term only. Accordingly, we split the dissipator as

$$\mathcal{L}_D = \mathcal{L}_{D_1} + \mathcal{L}_{D_2}. \tag{4.164}$$

The current I_2 from the level to the Γ_2 reservoir is determined by the temporal change of the electron density in the level due to the coupling to the reservoir Γ_2 only:

$$I_2 = -\frac{\mathrm{d}}{\mathrm{d}t}\langle a^\dagger a\rangle_{\Gamma_2} \tag{4.165}$$

$$= -\mathrm{Tr}\left(a^\dagger a\hat{\hat{\mathcal{L}}}_{D_2}\rho\right) \tag{4.166}$$

$$= -\langle I|\left[a^\dagger a, \hat{\mathcal{L}}_{D_2}\right]|\rho\rangle. \tag{4.167}$$

We leave it as an exercise to determine the steady-state current and show that in steady state the current is conserved $I_1 = -I_2$.

4.5.1.1 Generic Fermionic Hamiltonian with Many Levels

For the case of a central system consisting of N noninteracting fermionic levels with Hamiltonian

$$H = \sum_{nm} \varepsilon_{nm} a_n^\dagger a_m,$$

and dissipator (4.48), it is straightforward to show that the expressions (4.161) with (4.162) still hold, provided one takes ε, Γ_1, Γ_2 as matrices with elements ε_{nm}, Γ_{1nm}, Γ_{2nm}, as well as

$$\underline{\underline{a}}^\dagger = \left(a_1^\dagger, \ldots, a_N^\dagger, \widetilde{a}_1, \ldots, \widetilde{a}_N\right). \tag{4.168}$$

If, additionally, an interaction described by a Hamiltonian H_U is present in the central system, the corresponding contribution to the Lindblad operator being $\hat{\hat{\mathcal{L}}}_U \hat{\rho} = -i\left[H_U, \hat{\rho}\right]$, becomes in the superfermion representation [cf. (4.153)]

$$\hat{\mathcal{L}}_U |\rho\rangle = -i(H_U - \widetilde{H}_U)|\rho\rangle. \tag{4.169}$$

Example (Anderson impurity chain attached to reservoirs) As a simple example, one can consider a fermionic tight-binding chain consisting of N sites (spin is not indicated explicitly) in which the leftmost site $n = 1$ is attached to a reservoir injecting particles, Γ_2 with the only nonzero matrix element $\Gamma_{2\,1,1}$, and the rightmost site $n = N$ is attached to a reservoir removing particles, Γ_1 with the only nonzero matrix element $\Gamma_{1\,N,N}$. One can include a Hubbard interaction U on the central chain, so that the system describes a nonequilibrium Anderson impurity chain in which a current flows from left to right, see upper part of Fig. 4.6. The corresponding superfermion Hamiltonian describes two chains, one corresponding to the operators a_n, the other to the $\widetilde{a}_n$. The two chains are coupled by the Γ and have opposite sign of the single-particle parameters. The Γ_2 (Γ_1) term injects (removes) particles on both chains, so that the total particle number is not conserved (see lower part of Fig. 4.6). However, if one carries out a particle-hole transformation for the tilde particles $\widetilde{d}_n = \widetilde{a}_n^\dagger$, then the total particle number $\sum_{n=1}^N \left(a_n^\dagger a_n + \widetilde{d}_n^\dagger \widetilde{d}_n\right)$ is conserved.

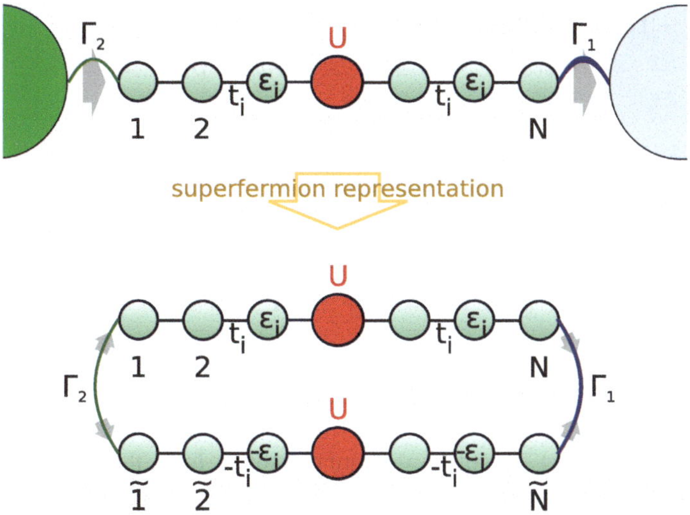

Fig. 4.6 (Top) Illustration of an Anderson impurity coupled to two reservoirs given by tight-binding chains with Lindblad drivings at the outermost sites. (Bottom) In the superfermion representation this maps onto to two chains, which are coupled via the Lindblad terms Γ_1 and Γ_2

4.6 Correlation Functions and Quantum Regression Theorem

Up to now we only discussed the time dependence of expectation values $\langle A(t)\rangle$. We now focus on two-time correlation functions $\langle A(t)B(t')\rangle$. The computation of such correlation or Green's functions is particularly important in the present treatment, since it enables us to combine the Lindblad approach with the framework of nonequilibrium Green's functions, as outlined below in more detail.

The time dependence of an operator A acting on the system only is given by

$$\langle A(t)\rangle = \text{Tr}A(t)\varrho = \text{Tr}_S\,\text{Tr}_R\,A\varrho(t) = \text{Tr}_S\,A\varrho_S(t), \tag{4.170}$$

with $\varrho_S = \text{Tr}_R\,\varrho$ the system reduced and ϱ the universe density matrix. Here, we have exploited the fact that the Heisenberg time evolution of an operator A has opposite sign with respect to the time evolution of ρ, the cyclic property of the trace, and that the reservoir trace can be "pulled over" the system operator A. Due to this, it is sufficient to know the time dependence of $\varrho_S(t)$, which is given by the Lindblad equation (4.41) as discussed up to now. However, for two-time correlation functions of system operators a knowledge of $\varrho_S(t)$ is no longer sufficient. Let us illustrate this for the following correlation function of two system operators A, B:

$$
\begin{aligned}
iG_{BA}(t_1+\tau,t_1) &= \mathrm{Tr}B(t_1+\tau)A(t_1)\varrho \\
&= \mathrm{Tr}e^{iH(t_1+\tau)}Be^{-iH(t_1+\tau)}e^{iHt_1}Ae^{-iHt_1}\varrho \\
&= \mathrm{Tr}Be^{-iH\tau}A\varrho(t_1)e^{iH\tau} = \mathrm{Tr}_S\, B\, \mathrm{Tr}_R\, e^{-iH\tau}A\varrho(t_1)e^{iH\tau}. \qquad (4.171)
\end{aligned}
$$

Now, since the Hamiltonian H acts on both system and reservoir, one cannot pull Tr_R over $e^{-iH\tau}$.

In order to make progress, let us introduce the following system operator

$$
A_S(\tau,t_1) := \mathrm{Tr}_R\, e^{-iH\tau}A\varrho(t_1)e^{iH\tau}, \qquad (4.172)
$$

in terms of which

$$
iG_{BA}(t_1+\tau,t_1) = \mathrm{Tr}_S\, BA_S(\tau,t_1). \qquad (4.173)
$$

Unfortunately, $A_S(\tau,t_1)$ cannot be determined solely from the knowledge of the reduced density matrix $\varrho_S(t_1)$. Fortunately, the so-called *quantum regression theorem* [85, 87] states that the time dependence of the operator $A_S(\tau,t_1)$ is governed by an equation of Lindblad type

$$
\frac{d}{d\tau}A_S(\tau,t_1) = \underline{\mathcal{L}}A_S(\tau,t_1), \qquad (4.174)
$$

provided that the same Markovian conditions as for ρ_S, (4.80), hold true:

$$
T_S \gg \Delta t \gg \tau_R. \qquad (4.175)
$$

This result combined with the initial ($\tau = 0$) condition

$$
A_S(0,t_1) = \mathrm{Tr}_R\, A\varrho(t_1) = A\varrho_S(t_1), \qquad (4.176)
$$

allows to determine an arbitrary operator $A_S(\tau,t_1)$, and thus any two-time correlation function $iG_{BA}(t+\tau,t_1)$. This works as follows:

(1) First calculate $\rho_S(t_1)$ from $\frac{d}{dt_1}\rho_S = \mathcal{L}\rho_S$ and a given initial condition. In particular, we are interested in the steady state case $t_1 \to \infty$, see below.
(2) Then compute the τ-time evolution of $A_S(\tau,t_1)$ from (4.174), with initial condition (4.176), taking t_1 as a fixed parameter.

In fact, for the case that A is a bosonic operator (or contains even products of fermionic creation/annihilation operators), $\underline{\mathcal{L}}$ and $\mathcal{L}$ from (4.41) coincide. For the case of operators containing odd products of fermions, which is relevant in evaluating single-particle Green's functions, there is an additional sign, [95], which we are going to discuss below.

The Quantum Regression Theorem (4.174) can be readily proven by repeating the steps of Sect. 4.4.4 whereby one takes instead of the universe density matrix $\rho(t)$, the quantity $[A\varrho(t_1)](t)$, where $\frac{d}{dt}[\cdots](t) = -i[H,[\cdots](t)]$, c.f. (4.53). Since the quantity we are looking for is precisely $A_S(t, t_1) = \mathrm{Tr}_R[A\varrho(t_1)](t)$ (cf. (4.172)), the procedure carried out to determine the time dependence of $\rho_S(t)$ ((4.54)) is precisely the same. As a result, one gets the same (4.174) with $\underline{\mathcal{L}} = \mathcal{L}$. For fermions, one should take care of the fact that the coupling Hamiltonian cannot be readily written in the form (4.52), since there are additional fermionic signs. In the end, this leads to a slightly different expression for $\underline{\mathcal{L}}$, which we are going to discuss below. See [95], Appendix B, for a complete treatment.

One should point out that the Lindblad time evolutions (4.174) and (4.41) are valid only in the positive direction of time. Inherently, this is connected to the Markov approximation and the dissipative dynamics. However, in the case of correlation functions we generally need to compute $iG_{BA}(t+\tau, t)$ for $\tau < 0$ as well. This can be achieved in two ways

- Instead of $iG_{BA}(t+\tau, t)$ one considers the complex conjugate

$$\begin{aligned} -iG^*_{BA}(t+\tau, t) &= \mathrm{Tr} A^\dagger(t) B^\dagger(t+\tau)\varrho \\ &= iG_{A^\dagger B^\dagger}(t, t+\tau), \end{aligned} \tag{4.177}$$

  which is in the proper time order since $t - (t+\tau) > 0$ for $\tau < 0$.
- Alternatively, with the cyclic invariance of the trace one has that

$$\begin{aligned} iG_{BA}(t+\tau, t) &= \mathrm{Tr} A(t)\varrho B(t+\tau) \\ &= \mathrm{Tr}_S\, A\, \mathrm{Tr}_R \left\{ e^{iH\tau}\varrho(t+\tau) B e^{-iH\tau} \right\} \\ &= \mathrm{Tr}_S\, A \widetilde{B}^\dagger_S(-\tau, t+\tau), \end{aligned} \tag{4.178}$$

  and the time evolution of $\widetilde{B}^\dagger_S(-\tau, t+\tau)$ is determined by (4.174) for $\tau < 0$.

4.6.1 Superfermion Representation

We now want to express a correlation function (4.173) in the superfermion formalism of Sect. 4.5. In this notation,

$$iG_{BA}(t_1+\tau, t_1) = \langle I|BA_S(\tau, t_1)|I\rangle \equiv \langle I|B|A_S(\tau, t_1)\rangle\!\rangle\ . \tag{4.179}$$

Here, the supervector $|A_S(\tau, t_1)\rangle\!\rangle$ has the properties

$$\frac{d}{d\tau}|A_S(\tau, t_1)\rangle\!\rangle = \hat{\mathcal{L}}|A_S(\tau, t_1)\rangle\!\rangle, \qquad |A_S(0, t_1)\rangle\!\rangle = A|\rho_S(t_1)\rangle\!\rangle, \tag{4.180}$$

provided A is a bosonic operator. This can be easily shown by using (4.174), (4.176), and (4.144), and proceeding like for (4.145, 4.148, 4.151). For fermionic operators the derivation is somewhat more tricky, but in the end one obtains effectively the same expression (4.180) with the same $\hat{\mathcal{L}}$, despite of the fact that in (4.174) $\hat{\hat{\underline{\mathcal{L}}}}$ differs from $\hat{\hat{\mathcal{L}}}$. This is discussed in the next section.

Note that the expression (4.179) is valid for $\tau > 0$ only. For negative τ one should use (4.177) or (4.178).

4.6.2 *Fermionic Operators*

As mentioned above, special care has to be taken for the case of fermionic operators since their expression in terms of tensor products is not trivial. We here only sketch the issue and refer to [95], Appendix B, for a complete treatment. The system operator A of Sect. 4.6 has in fact the form $A = I_R \otimes A_S$. On the other hand, a single-particle fermionic operator C for the system does not have this form since it anti-commutes with reservoir states. Consider for instance the product state $|\psi\rangle = |R\rangle \otimes |S\rangle$. Here,[16]

$$C|\psi\rangle = (-1)^{N_R}|R\rangle \otimes C_S|S\rangle \,, \tag{4.181}$$

with N_R the number of fermions in state $|R\rangle$. Therefore, one must include these phase factors in the definition of the tensor product operators, leading to

$$C = (-1)^{N_R} \otimes C_S. \tag{4.182}$$

When carrying out the microscopic derivation of Sect. 4.4.4, one finds that these sign factors cancel away in the case of the Lindblad equation for the density matrix ρ, while for correlation functions they do matter.

Following the treatment of [95], Appendix B, one obtains that a fermionic operator $C_S(\tau, t_1)$ defined similarly to (4.172) obeys

$$\frac{d}{d\tau} C_S(\tau, t_1) = \hat{\hat{\underline{\mathcal{L}}}} C_S(\tau, t_1) \,, \tag{4.183}$$

with

$$\hat{\hat{\underline{\mathcal{L}}}} C_S := -i\,[H, C_S] + \sum_{\alpha\beta} \gamma_{\alpha\beta} \left(\eta S_\alpha C_S S_\beta^\dagger - \frac{1}{2} \left\{ S_\beta^\dagger S_\alpha, C_S \right\} \right). \tag{4.184}$$

The additional sign factor η (possibly) distinguishes this result from (4.41) and is equal to -1 if C_S and S_α both contain an odd number of fermionic operators, and $+1$ otherwise.

---

[16] C_S is the same as C but acting on the system Hilbert space only.

Nevertheless, the pleasant aspect is that in the superfermion representation of Sect. 4.5, this sign η cancels out again. Therefore, in the superfermion representation, the vector $|C_S(\tau, t_1)\rangle$ associated to $C_S(\tau, t_1)$ obeys an equation like (4.155)

$$\frac{d}{d\tau}|C_S(\tau, t_1)\rangle = \hat{\mathcal{L}}|C_S(\tau, t_1)\rangle, \tag{4.185}$$

with the same $\hat{\mathcal{L}}$ (4.153), (4.154).

4.7 Nonequilibrium Green's Functions

Nonequilibrium Green's functions have been treated in detail in the previous two lectures, so here we are simply going to summarize the parts which are most relevant for the present treatment. Again we specialize to the case of a fermionic system. We refer to these lectures and to previous literature (see, e.g. [1, 2, 96]).

As introduced by Kadanoff, Baym and Keldysh, a modified time contour ordering allows one to formulate a systematic Green's function formalism analogous to the equilibrium case. In contrast to equilibrium, the system states at $t \to \pm\infty$ are no longer equivalent and thus the only reference point is the infinite past.[17] Only there one can assume that the system is in a noninteracting initial state necessary in order to apply Wick's theorem. Therefore, instead of time-ordered expectation values as in equilibrium, one has to consider contour-ordered ones. Different contour orders exist and we focus here only on the Keldysh contour, as sketched in Fig. 4.7. Here, the Matsubara branch, accounting for initial correlations, is neglected and the contour extends until $t \to -\infty$. This is justified when considering steady states or even when carrying out time evolutions starting from a steady state.[18] An example for a two-time correlation function is depicted in Fig. 4.7, which demonstrates that the contour-ordering of times generally differs from the ordinary time-ordering. When denoting contour times by $\tau_{A/B}$ and "standard" times by $t_{A/B}$, one can write contour-ordered two-time Green's functions in the following way

$$G(\tau_A, \tau_B) \Rightarrow \hat{G}(t_A, t_B) = \begin{pmatrix} G_T(t_A, t_B) & G^<(t_A, t_B) \\ G^>(t_A, t_B) & G_{\tilde{T}}(t_A, t_B) \end{pmatrix}. \tag{4.186}$$

It is convenient to employ a matrix structure, which contains all the possible orderings of the two time variables $t_{A/B}$ on the lower and on the upper contour. $G_T(t_A, t_B)$ ($G_{\tilde{T}}(t_A, t_B)$) is the time (anti-time) ordered Green's function, which corresponds to the case that t_A and t_B are both on the upper (lower) contour. The lesser (greater)

[17] In case of the L-shaped Kadanoff-Baym contour this starting point is on the imaginary-time Matsubara branch, i.e. corresponds to a thermal initial state.

[18] In principle, one could avoid the Matsubara branch altogether by designing a Hamiltonian whose steady state is the required initial state.

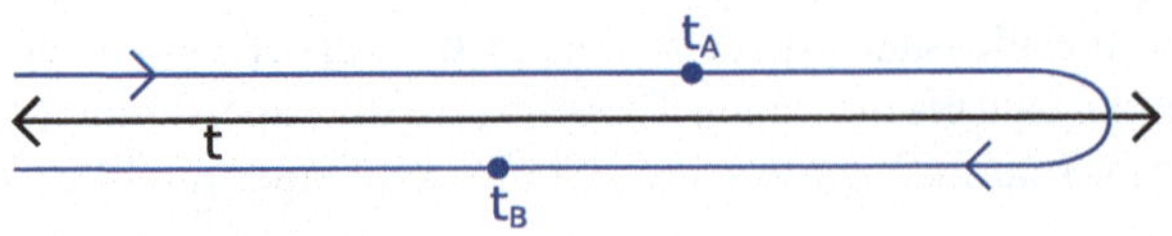

Fig. 4.7 Sketch of the Keldysh contour with an upper and a lower branch, both extending to $-\infty$. Depicted is the example of a "lesser" two-time function (e.g. $G^<(t_A, t_B)$), in which t_A is before t_B in terms of the contour ordering

Green's function $G^<(t_A, t_B)$ ($G^>(t_A, t_B)$) refers to the mixed cases, with one time variable on the upper and one on the lower time contour.

The matrix form stated above contains redundant information and it is thus convenient to employ a transformation [2]

$$\underline{G} = L\sigma_3 \hat{G} L^\dagger = \begin{pmatrix} G_R & G_K \\ 0 & G_A \end{pmatrix}, \tag{4.187}$$

to the so-called Keldysh space. The retarded (G_R), advanced (G_A) and Keldysh (G_K) Green's functions are hereby defined as:

$$\begin{aligned} G_R(1,2) &= -i\Theta(t_1 - t_2)\langle\{c(1), c^\dagger(2)\}\rangle, \\ G_A(1,2) &= G_R(2,1)^\dagger, \\ G_K(1,2) &= -i\langle[c(1), c^\dagger(2)]\rangle. \end{aligned} \tag{4.188}$$

The matrix form (4.187), which we shall indicate by an underscore "_", is very useful since essentially the full perturbation theory and Feynman diagrams developed for equilibrium is also applicable in the nonequilibrium case, whereby all scalar expressions for the Green's functions have to replaced by analogous matrix expressions.

Besides matrix products, one also needs to compute inverses $\underline{F}^{-1}$ of two-point Keldysh objects $\underline{F}$. This is given in terms of the Langreth rules, by

$$\underline{F} = \begin{pmatrix} F_R & F_K \\ 0 & F_A \end{pmatrix} \quad \rightarrow \quad \underline{F}^{-1} = \begin{pmatrix} F_R^{-1} & -F_R^{-1} F_K F_A^{-1} \\ 0 & F_A^{-1} \end{pmatrix}, \tag{4.189}$$

whereby the individual objects $F_R, F_K, \ldots$ can also be matrices in site and/or spin indices. Clearly, retarded objects transform in a simple manner and only the Keldysh part is more involved

$$\begin{aligned} (\underline{F}^{-1})_R &= (F_R)^{-1}, \\ (\underline{F}^{-1})_K &= -F_R^{-1} F_K F_A^{-1}. \end{aligned} \tag{4.190}$$

4.7.1 Anderson Impurity Model

As mentioned at the beginning, we are particularly interested in the nonequilibrium Anderson impurity model, which is described by the Hamiltonian

$$\begin{aligned} H &= H_C + H_R + V, \\ H_C &= \varepsilon \sum_{\sigma} c^{\dagger}_{0\sigma} c_{0\sigma} + U n_{0\uparrow} n_{0\downarrow}, \\ H_R &= \sum_{\sigma, p\neq 0} \varepsilon_p c^{\dagger}_{p\sigma} c_{p\sigma}, \\ V &= \sum_{\sigma, p\neq 0} v_p c^{\dagger}_{p\sigma} c_{0\sigma} + \text{h.c.}, \end{aligned} \tag{4.191}$$

with H_C the impurity, H_R the reservoir and V the coupling Hamiltonian.[19] For the reservoir we consider the case of two leads denoted by $+$ and $-$, corresponding to $p > 0$ and $p < 0$ in (4.191), with different chemical potentials (μ_+, μ_-) and temperatures (T_+, T_-), see left side of Fig. 4.8. The unperturbed Hamiltonian

$$H_0 = H_- + H_+ + H_{C0}. \tag{4.192}$$

corresponds to the decoupled system without interaction and we consider as perturbation the hybridizations v_p and the interaction U. At $t_0 \to -\infty$ the system is prepared in an eigenstate of H_0, i.e. the three regions are separately in equilibrium with their respective chemical potentials and temperatures, and the perturbation is then switched on. For $t - t_0 \to \infty$ the system reaches the steady state of the full Hamiltonian (4.191). In the steady state one can assume that time translational invariance applies, so that Green's functions can be written in the frequency domain:

$$\underline{G}(t_1 - t_2) \to \underline{G}(\omega) \,. \tag{4.193}$$

From now on we assume that all Green's functions are ω-dependent and omit the argument for the sake of simplicity.

Let us start with the noninteracting case $U = 0$, so that the perturbation is only given by the hybridizations to the leads. For this case the exact Dyson equation reads

$$\underline{G} = \underline{g} + \underline{g}\, \underline{V}\, \underline{G}. \tag{4.194}$$

The equation is analogous to the equilibrium case, the difference being that every object has a 2×2 matrix structure in Keldysh space, in addition to level and/or

[19] Similarly, one could also generalize the steps outlined below to the situation of a central system containing a small number of interacting sites, see Fig. 4.1, by a suitable matrix notation.

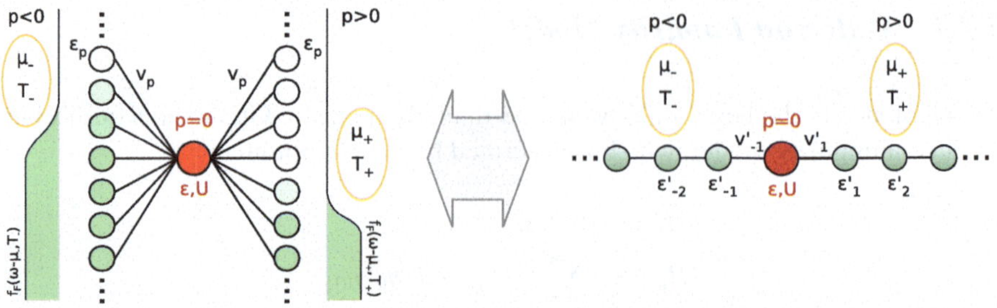

Fig. 4.8 (Left) Sketch of the nonequilibrium Anderson impurity model as defined in (4.191). The two reservoirs $p < 0$ and $p > 0$ consist of an infinite number of levels ε_p, which are (at $t_0 \to -\infty$) filled according to the Fermi-Dirac distributions $f_F(\omega - \mu_\pm, T_\pm)$. This is the "star" representation. (Right) Equivalent "chain" representation, with two semi-infinite chains representing the reservoirs

spin indices. $\underline{G}$ is the full Green's function, $\underline{g}$ is the Green's function of the isolated regions ($v_p = 0$) and $\underline{V}$ is the hybridization, which is diagonal in Keldysh space:

$$\underline{V}_{0p} = \underline{V}_{p0} = \begin{pmatrix} v_p & 0 \\ 0 & v_p \end{pmatrix}. \tag{4.195}$$

In principle, the full matrices in (4.194) can be inverted with the help of (4.189). It is convenient to write them explicitly in terms of their components

$$\begin{aligned} \underline{G}_{00} &= \underline{g}_{00} + \underline{g}_{00} \sum_p \underline{V}_{0p} \underline{G}_{p0}, \\ \underline{G}_{p0} &= \underbrace{\underline{g}_{p0}}_{0} + \sum_{p'} \underline{g}_{pp'} \underline{V}_{p'0} \underline{G}_{00}. \end{aligned} \tag{4.196}$$

Here, we made use of the fact that $\underline{g}_{pp'}$ does not have off-diagonal components linking the initially decoupled regions. On the whole, one can write Dyson's equation as

$$\underline{G}_{00} = \underline{g}_{00} + \underline{g}_{00}\, \underline{\Delta}\, \underline{G}_{00}, \tag{4.197}$$

with the bath *hybridization function* defined as

$$\underline{\Delta} = \sum_{p,p'} \underline{V}_{0p} \underline{g}_{pp'} \underline{V}_{p'0}\,. \tag{4.198}$$

As usual, the solution of (4.197) is obtained by

$$\underline{G}_{00} = \left(\underline{g}_{00}^{-1} - \underline{\Delta} \right)^{-1}, \tag{4.199}$$

whereby one has to take the Langreth rules (4.189) into account, in order to invert the 2×2 Keldysh objects.

The reservoir Green's functions $\underline{g}_{pp'}$ are known analytically, since they correspond to a noninteracting system in equilibrium.[20] For a reservoir Hamiltonian as specified in (4.191), which is diagonal in the p operators, the retarded part is given by

$$g_{Rpp'}(\omega) = \delta_{pp'} \left(\omega - \varepsilon_p + i0^+\right)^{-1} . \tag{4.200}$$

Of course, other choices of the reservoir are possible as well, e.g. a "chain" instead of a "star" representation, see Fig. 4.8. In the latter case, only one site of each lead, e.g. $g_{R11}(\omega)$ and $g_{R-1-1}(\omega)$, would couple to the central system. Notice that such "surface" Green's function of a noninteracting semi-infinite tight-binding chain can be determined analytically, cf. [97].

In equilibrium, the Keldysh and the retarded Green's functions are not independent but linked via the so-called *fluctuation dissipation theorem*:

$$\begin{aligned} g_{Kpp}(\omega) &= \left(g_{Rpp}(\omega) - g_{App}(\omega)\right) s_p(\omega), \\ s_p(\omega) &= 1 - 2 f_F(\omega - \mu_p, T_p), \end{aligned} \tag{4.201}$$

with $f_F(\omega - \mu_p, T_p)$ the Fermi-Dirac distribution. For the nonequilibrium case, the Keldysh and the retarded component are independent functions and both of them must be considered explicitly.

As in equilibrium, the solution of the interacting problem $U \neq 0$ poses the main challenge. A couple of different approaches are discussed in the next section. Here, let us focus on the general properties. As usual, the contribution from U can be encoded in terms of the self energy $\underline{\Sigma}(\omega)$, which is also a 2×2 Keldysh object in nonequilibrium. In terms of site indices pp', $\underline{\Sigma}(\omega)$ is only nonzero when an interaction term is present in the Hamiltonian at p and p'. Therefore, in the single impurity case considered here, the self energy has only contributions on the impurity site. In this way, (4.199) is modified to

$$\underline{G}_{00} = \left(\underline{g}_{00}^{-1} - \underline{\Delta} - \underline{\Sigma}_{00}\right)^{-1} . \tag{4.202}$$

Once $\underline{\Sigma}_{00}$ is known, all single particle quantities of interest can be computed.[21] The possibly spin-dependent particle density on the impurity site, for instance, is given in terms of the Keldysh Green's function by

$$n = \frac{1}{2} - \frac{i}{4\pi} \int G_{K00}(\omega) \mathrm{d}\omega . \tag{4.203}$$

---

[20] Note that $\underline{g}$ refers to the initially decoupled situation.

[21] We focus here on the impurity, but also reservoir properties are accessible via Dyson's equation.

The current from the reservoir to the impurity is determined in terms of G_{Kp0} leading to the Meir-Wingreen formula: [98]

$$j = \frac{i}{2\pi}\int_{-\infty}^{\infty} \mathrm{d}\omega \left(\left[\gamma_-(\omega) - \gamma_+(\omega)\right] G_{00}^{<}(\omega) \right.$$
$$\left. + \left[f_{F-}(\omega)\gamma_-(\omega) - f_{F+}(\omega)\gamma_+(\omega)\right]\left[G_{R00}(\omega) - G_{A00}(\omega)\right]\right), \quad (4.204)$$

with $f_{F\pm}$ the Fermi functions of the left (-) and right (+) reservoir, and $\gamma_\pm(\omega) = -2\Im\mathrm{m}\,\{\Delta_{R\pm}(\omega)\}$ accounts for the coupling strength to and the DOS of each lead.

4.8 Nonequilibrium Impurity Problems

The manybody solution of nonequilibrium impurity problems, as defined by (4.191), is an active area of research and numerous different approaches were devised in recent years. Here, we want to give only a brief overview over some of them and then focus on solution strategies based on a combination of nonequilibrium Green's functions and Lindblad equations, which is the topic of the present lecture.

Powerful numerical methods for the solution of equilibrium impurity models are for instance exact diagonalization (ED), quantum Monte Carlo (QMC), matrix product states (MPS) and numerical renormalization group (NRG). Except for action-based QMC solvers, the common solution strategy is to replace the exact hybridization function $\underline{\Delta}(\omega)$ by an approximate one, corresponding to a finite size system which can be solved precisely by numerical techniques (see, e.g., Fig. 4.9).

The key point is always that the influence of the leads is completely determined by $\underline{\Delta}(\omega)$. In other words, the self energy $\underline{\Sigma}(\omega)$ depends solely on ε, U and $\underline{\Delta}(\omega)$, but not on other details of the reservoir. This means that different representations of the reservoir, for instance a chain or a star geometry, which yield the same $\underline{\Delta}(\omega)$ are equivalent on the level of impurity properties. Both of them result in the same $\underline{G}_{00}$

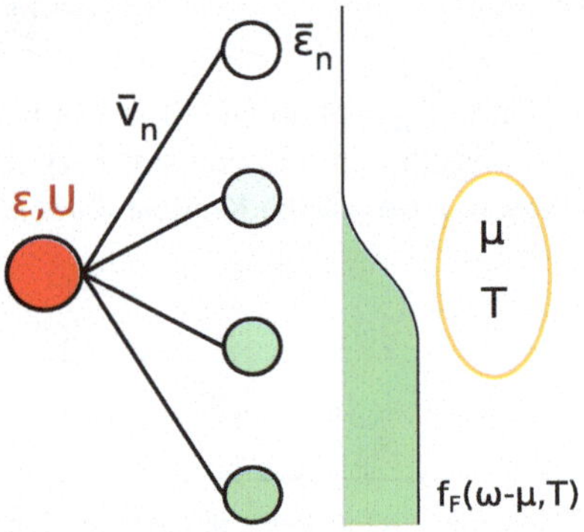

Fig. 4.9 Exact diagonalization approach as used for equilibrium situations. Instead of the exact system, Fig. 4.8 with a single reservoir, e.g. $p > 0$ only, a finite size problem consisting of the impurity and a small number of levels $\bar{\varepsilon}_n$ is solved

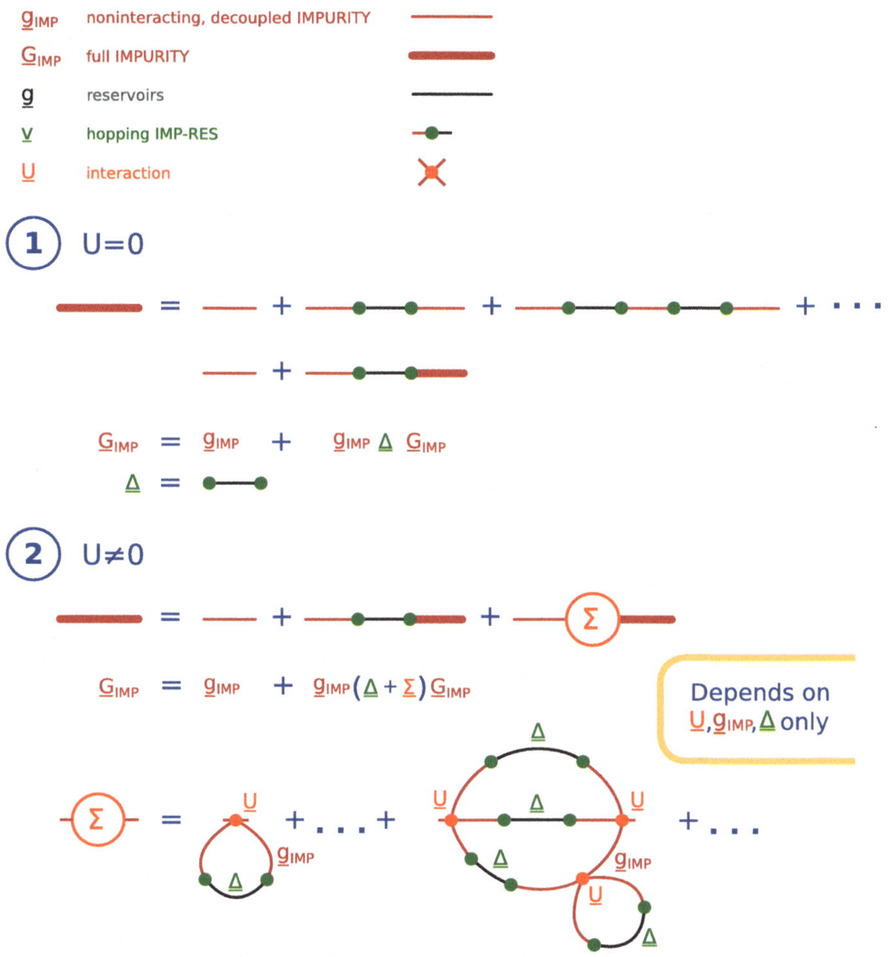

Fig. 4.10 Sketch of the diagrammatic proof that correlation functions on the impurity site are fully determined by the hybridization function $\underline{\Delta}(\omega)$, and the impurity terms U and $\underline{g}_{\mathrm{IMP}}$. Other details of the bath are irrelevant, e.g. whether one considers a "star" or "chain" representation, cf. Fig. 4.8. The bath can be also represented by a mixed auxiliary system consisting of orbitals and Lindblad terms, such as a buffer layer (see Sect. 4.8.1) or by a more generic one within the Auxiliary Master Equation Approach [21, 63], as depicted in Fig. 4.13, see Sect. 4.9

and $\underline{\Sigma}_{00}$. Of course, this fact holds in nonequilibrium as well, see Fig. 4.10. Within NRG, this fact is exploited to justify the Wilson chain [6].

In equilibrium one exploits this to replace the dense reservoir by an auxiliary reservoir with a small number of levels only and different parameters $\bar{\varepsilon}_n$, $\bar{v}_n$, see Fig. 4.9. Here, the parameters are determined (fitted) in order to provide the best representation of the bath hybridization function $\Delta(i\omega_\lambda)$ in Matsubara frequency space. This is the exact-diagonalization based impurity solver [8, 9], widely used for DMFT.

Out of equilibrium this does not work since a finite size reservoir cannot provide dissipation and thus a steady state situation can never be reached in the time evolution. Instead, such a system exhibits oscillating dynamics. Here, we want to consider closely related approaches, in which the reservoir is modeled by a small number of levels which are additionally coupled to Markovian environments. Such auxiliary systems are governed by a Lindblad equation, which we discussed earlier. The key advantage is that these reservoir representations exhibit dissipative dynamics and truly represent nonequilibrium impurity systems.

4.8.1 Buffer Layer Approach

In the so-called buffer layer approach, see e.g. [65], one considers a certain number N_B of bath levels coupled to the impurity site, similar to the original Hamiltonian (4.191) but with N_B finite. To "compensate" for the missing part of the infinite reservoir one additionally couples the bath sites to Markovian environments, see also Fig. 4.11. In this way one is able to achieve a continuous DOS in the auxiliary system, appropriate for a nonequilibrium situation.

If one assumes for the Markovian environments an infinite bandwidth and energy-independent occupations n_n, the auxiliary system can be exactly written in terms of a Lindblad equation, as previously discussed

$$\dot{\rho} = \mathcal{L}_H \rho + \mathcal{L}_D \rho,$$
$$\mathcal{L}_D \rho = 2 \sum_{n=1}^{N_B} \Big(\Gamma_{1n} \big(d_n \rho d_n^\dagger - \frac{1}{2} \{ d_n^\dagger d_n, \rho \} \big) + \Gamma_{2n} \big(d_n^\dagger \rho d_n - \frac{1}{2} \{ d_n d_n^\dagger, \rho \} \big) \Big), \quad (4.205)$$

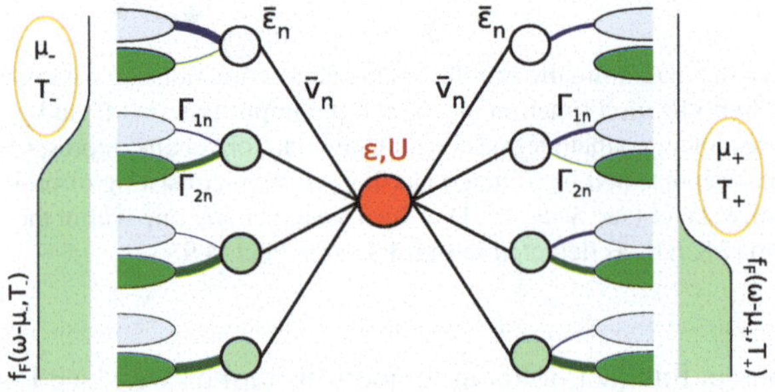

Fig. 4.11 Buffer layer approach: The nonequilibrium impurity model Fig. 4.8 is replaced by a finite number of levels $\bar{\varepsilon}_n$ which are additionally coupled to Markovian environments. The appropriate filling $n_n = f_F(\bar{\varepsilon}_n - \mu_\pm, T_\pm)$ is achieved by suitable coupling constants Γ_{1n} and Γ_{2n} to the empty and filled Markovian environments, (4.206). The resulting finite size open quantum system is governed by a Lindblad equation (4.205) and represents a true nonequilibrium model

where ρ represents the density matrix of the open system consisting of impurity plus level sites with corresponding operators d_n. There are two types of Markovian environments, one completely empty ($\mu \to -\infty$) and one completely filled ($\mu \to +\infty$). The coefficients Γ_{1n} determine the couplings to the empty environment, and Γ_{2n} the couplings to the filled one. One can choose them in the following way:

$$\Gamma_{1n} = \bar{M}_n(1 - n_n),$$
$$\Gamma_{2n} = \bar{M}_n n_n. \tag{4.206}$$

Here, $\bar{M}_n$ determines the coupling strength of the n-th level to the two Markovian environments, and n_n refers to its desired occupation.

We now evaluate the corresponding auxiliary bath hybridization function $\underline{\Delta}^{Aux}$ at the impurity site. For this one should first determine the noninteracting Green's function and then use (4.199). The expression for the noninteracting Green's function of an open lattice system described by (4.161), (4.162) is evaluated in Sect. 4.9.1.1, and the expression for the Green's function matrices is given in (4.222).

For the present case it is more convenient to use (4.198) in terms of the local Green's function $\underline{g}_{nn} = \underline{g}_n$ of the n-th isolated level plus Markovian reservoir, i.e. decoupled from the impurity site. These can be determined by using (4.222) for a single site, leading to

$$g_{Rn} = (\omega - \bar{\varepsilon}_n + i\Gamma_{+n})^{-1},$$
$$g_{Kn} = 2i g_{Rn}(\Gamma_{2n} - \Gamma_{1n}) g^*_{Rn} = \frac{2i(\Gamma_{2n} - \Gamma_{1n})}{(\omega - \varepsilon_n)^2 + \Gamma^2_{+n}}. \tag{4.207}$$

With (4.198), the auxiliary hybridization function on the impurity site is given by

$$\Delta_R^{Aux} = \sum_n \bar{v}_n^2 g_{Rn},$$
$$\Delta_K^{Aux} = \sum_n \bar{v}_n^2 g_{Kn}. \tag{4.208}$$

Now, the goal is to approximate the physical Δ_R, Δ_K as accurately as possible by Δ_R^{Aux} and Δ_K^{Aux}. Due to the Kramers-Kronig relation between the imaginary and the real part of retarded functions, it is sufficient to consider only the imaginary part of Δ_R, and Δ_K is itself purely imaginary. The bath spectral function, determining the DOS of the auxiliary reservoir, is given by

$$A_\Delta^{Aux}(\omega) \equiv -\frac{1}{\pi}\Im\mathrm{m}\left\{\Delta_R^{Aux}(\omega)\right\}$$
$$= \sum_n \bar{v}_n^2 \delta_{\Gamma_{+n}}(\omega - \bar{\varepsilon}_n), \tag{4.209}$$

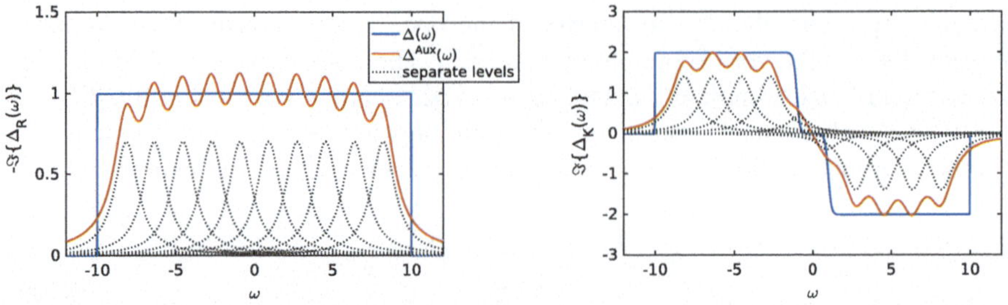

Fig. 4.12 Sketch of $\underline{\Delta}^{Aux}(\omega)$ in the buffer layer approach, with parameters chosen according to (4.211) and (4.212), and description in the text. The separate levels produce Lorentzian curves. For illustrative purposes we choose $\Gamma_{+n} = \delta_n/2$, see e.g. [95] for a detailed discussion

with the Lorentzians

$$\delta_\Gamma(\omega) \equiv \frac{1}{\pi}\frac{\Gamma}{\omega^2 + \Gamma^2}. \tag{4.210}$$

Therefore, a given physical bath spectral function $A_\Delta(\omega) = -1/\pi \Im\text{m}\{\Delta_R(\omega)\}$ is approximated by a superposition of Lorentz curves, as sketched in Fig. 4.12. For equidistant levels with energies $\bar{\varepsilon}_n$ the level spacing is given by

$$\delta_n = \frac{W}{N_B}, \tag{4.211}$$

with W the bandwidth. The width of the Lorentzians (4.210) is given by Γ_{+n}, and one should choose

$$\Gamma_{+n} \approx \delta_n, \tag{4.212}$$

in order to achieve a smooth and non-peaked $A_\Delta^{Aux}(\omega)$, which reproduces features in $A_\Delta(\omega)$ properly. The hoppings $\bar{v}_n$ are then adjusted in such a way that the local density of states is correctly reproduced, and normalization requires that $\sum_n \bar{v}_n^2 = 1$. In the hypothetical $N_B \to \infty$ limit, one recovers the exact result $A_\Delta^{Aux}(\omega) \to A_\Delta(\omega)$. See [95] for a further discussion of these aspects.

Up to now we only made use of $\bar{v}_n$ and Γ_{+n}, but $\Gamma_{-n} = \Gamma_{2n} - \Gamma_{1n}$ was not determined. This remaining degree of freedom amounts to specifying the filling of each Lorentz peak $\delta_{\Gamma_{+n}}(\omega)$, and thus to adjusting the Keldysh component Δ_K^{Aux}. From (4.201) one knows that the latter is related to the spectral function via

$$\Delta_K(\omega) = -2\pi i A_\Delta(\omega)s(\omega), \tag{4.213}$$

with $s(\omega) = 1 - 2n(\omega)$ the particular equilibrium or nonequilibrium occupation. From (4.207) and (4.208) we have for the auxiliary system

$$\Delta_K^{Aux}(\omega) = 2i\pi \sum_n \bar{v}_n^2 \frac{\Gamma_{-n}}{\Gamma_{+n}} \delta_{\Gamma_+}(\omega - \bar{\varepsilon}_n), \tag{4.214}$$

which suggests to determine Γ_{-n} via

$$\frac{\Gamma_{-n}}{\Gamma_{+n}} = -s(\bar{\varepsilon}_n), \tag{4.215}$$

in order to achieve the correct $N_B \to \infty$ limit. Furthermore, when inserting (4.206) into this expression we find

$$\frac{\Gamma_{-n}}{\Gamma_{+n}} = 2n_n - 1\,, \tag{4.216}$$

and thus $s(\bar{\varepsilon}_n) = 1 - 2n_n$, as desired. For an exemplary plot of Δ_R^{Aux} and Δ_K^{Aux} with the buffer layer idea see Fig. 4.12.

4.8.2 Finite Size Lindblad Impurity Problem

As sketched above, in the buffer layer approach one is able to approximate the original nonequilibrium impurity model by an auxiliary one, with a finite number N_B of levels only. Most importantly, these levels are coupled to additional Markovian reservoirs in order to obtain dissipation. Once this mapping to a finite size Lindblad model has been achieved, one can solve the auxiliary manybody problem with $U \neq 0$. This is much simpler than in the original model because one only has to deal with a finite many-body Hilbert space. Appropriate methods for this, such as Lanczos ED or MPS are discussed below.

On the whole, from solving the $U \neq 0$ Lindblad model one obtains the interacting Green's function $\underline{G}^{Aux}$ on the impurity site. From the discussion at the beginning of this section it is clear that the mismatch between the auxiliary $\underline{G}^{Aux}$ and the exact $\underline{G}$ of the original model depends solely on $\|\underline{\Delta}^{Aux} - \underline{\Delta}\|$. This difference can be reduced by increasing N_B. However, a larger number of bath sites comes at the price of a drastically increased effort to address the manybody solution. For example, methods such as Lanczos ED scale exponentially in N_B. Therefore, it is highly desirable to find a mapping procedure which yields a good accuracy $\|\underline{\Delta}^{Aux} - \underline{\Delta}\| \ll \|\underline{\Delta}\|$ already for modest values of N_B. This is provided by the Auxiliary Master Equation Approach [21, 63, 78] which we present in the next section.

4.9 Auxiliary Master Equation Approach

The key idea of the auxiliary master equation approach (AMEA), introduced in [21, 63], is to make optimal use of all available parameters in the finite size Lindblad impurity model. In this way, it is possible to achieve already for small values of $N_B \approx O(10)$ a very good accuracy $\|\underline{\Delta}^{Aux} - \underline{\Delta}\| \ll \|\underline{\Delta}\|$. For this purpose we consider the *most general* quadratic Lindblad dissipator with one impurity and N_B bath levels

$$\mathcal{L}_D\rho = 2\sum_{n,m=1}^{N_B}\left(\Gamma_{1mn}\big(d_n\rho d_m^\dagger - \frac{1}{2}\{d_m^\dagger d_n, \rho\}\big) + \Gamma_{2mn}\big(d_m^\dagger \rho d_n - \frac{1}{2}\{d_n d_m^\dagger, \rho\}\big)\right). \tag{4.217}$$

Note here the important aspect that the coupling matrices Γ_{1nm} and Γ_{2nm} to the Markovian environments are *non diagonal*, in contrast to (4.205), i.e. every possible coupling is included. The unitary part of the Lindblad equation can be chosen to be sparse, even in the most general case, since one always has the freedom to perform unitary transformations among bath sites only, since it does not affect $\underline{\Delta}^{Aux}$ on the impurity site. Therefore, the N_B bath sites can be assumed to be either in a chain or star geometry. The former case is schematically depicted in Fig. 4.13. In the latter case, in addition to Γ_{1nm} and Γ_{2nm}, we have the Hamiltonian parameters $\bar{\varepsilon}_n$ and $\bar{v}_n$. In AMEA all these parameters are optimized in a fit procedure by minimizing the difference between $\underline{\Delta}^{Aux}$ and $\underline{\Delta}$. We again stress that *the better* $\underline{\Delta}^{Aux}$ *approaches* $\underline{\Delta}$, *the better the accuracy of the impurity solver is.* In order to carry out the fit, we define the cost function

$$\chi(\bar{\varepsilon}, \bar{v}, \Gamma) = \int d\omega \|\Im\text{m}\left\{\underline{\Delta}^{Aux}(\omega) - \underline{\Delta}(\omega)\right\}\|^2, \tag{4.218}$$

and minimize it with respect to $\{\Gamma_{1nm}, \Gamma_{2nm}, \bar{\varepsilon}_n, \bar{v}_n\}$. On the whole, this amounts to solving a multi-dimensional optimization problem with $O(N_B^2)$ parameters. Even though this procedure is more involved than the simple choice of $\{\Gamma_{1nn}, \Gamma_{2nn}, \bar{\varepsilon}_n, \bar{v}_n\}$

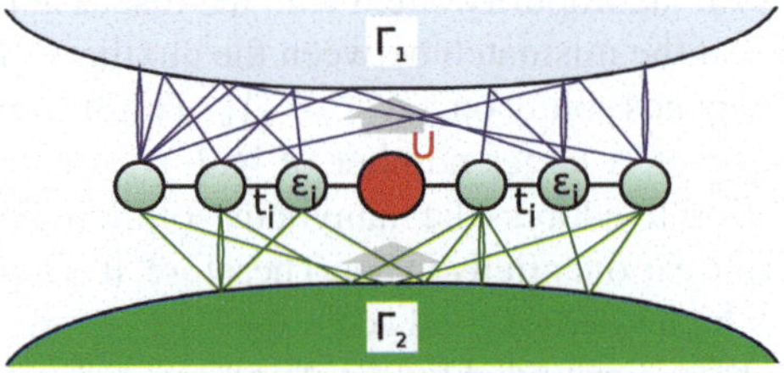

Fig. 4.13 Auxiliary master equation approach: The reservoirs are represented by a finite number of levels coupled to two Markovian environments. In contrast to the buffer layer approach, Fig. 4.11, all possible couplings are allowed, so that the coupling matrices Γ_{1nm} and Γ_{2nm} are *non diagonal*. Again, the system is governed by a Lindblad equation (4.217)

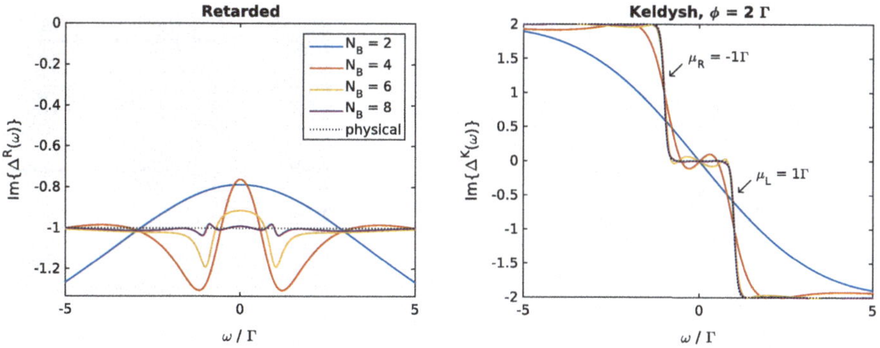

Fig. 4.14 Results for $\underline{\Delta}^{Aux}(\omega)$ in the auxiliary master equation approach, for a physical $\underline{\Delta}(\omega)$ similar to the one in Fig. 4.12. Already for real, dense Γ_{1nm} and Γ_{2nm} matrices, a rapid convergence with increasing N_B is observed, see also [63, 71, 78]. The bias voltage between left and right lead is denoted by ϕ, and Γ refers to the hybridization strength of the leads

in the buffer layer idea discussed above, the key aspect is that one achieves here an exponential convergence of $\underline{\Delta}^{Aux}$ towards $\underline{\Delta}$ with increasing N_B, see also [71]. Exemplary results for this fitting procedure are depicted in Fig. 4.14. More details can be found in [71].

Once the parameters of the auxiliary system are fitted, the interacting Lindblad equation must be solved. Up to now, we employed for this two different strategies based on ED and MPS, respectively, see [63] and [78] for details. The former allows us to consider all possible couplings so that modest values of $N_B = 6$ are sufficient for an accurate representation of the reservoirs, see Fig. 4.14. In the latter case we restricted the fit to sparse Γ_{1nm} and Γ_{2nm} matrices, in order to be able to apply efficient MPS techniques. The restriction results in not as optimal fits as with dense Γ_{1nm} and Γ_{2nm}, but, due to MPS much larger system sizes of $N_B = 15 - 20$ are possible, which outweighs and yields a significantly improved accuracy compared to the ED solver.

In the figures below, we present exemplary results for the ED and the MPS approach. In Figs. 4.15 and 4.16 ED results for a semi-circular lead DOS are shown. With the MPS approach we focused on the low-bias regime and considered a wide band model for the leads, as plotted in Figs. 4.12 and 4.14. Very accurate spectral functions could be obtained and it was possible for us to resolve the splitting of the Kondo peak with increasing bias voltage $\phi = \mu_L - \mu_R$ in detail, see Figs. 4.17, 4.18 and 4.19.

4.9.1 *Evaluation of Steady State Green's Functions*

We now focus on the computation of the steady state Green's functions in the auxiliary Lindblad system. At first we consider the noninteracting case, for which compact expressions are derived. These are crucial for AMEA in order to efficiently compute

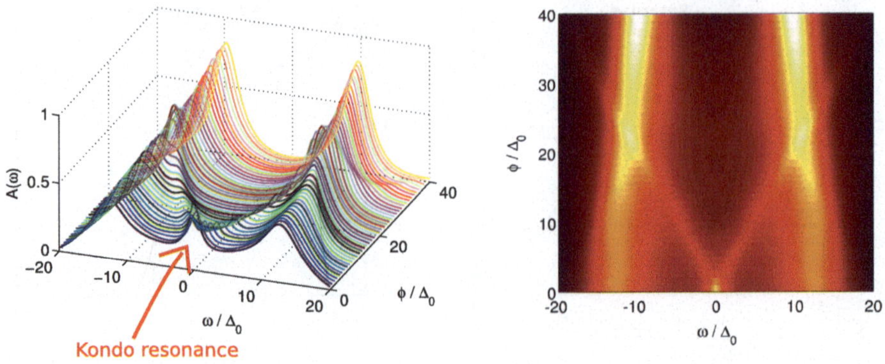

Fig. 4.15 ED results for the evolution of the impurity spectral function $A(\omega)$ with increasing bias voltage ϕ. In the equilibrium limit $\phi = 0$, a distinct Kondo peak and two Hubbard bands are clearly visible. The former splits upon increasing ϕ into two weak excitations, which are located at the positions of the chemical potentials $\mu_\pm = \pm\phi/2$. Results taken from [63], for an interaction strength $U = 20\,\Delta_0$, with Δ_0 half the hybridization strength of the leads

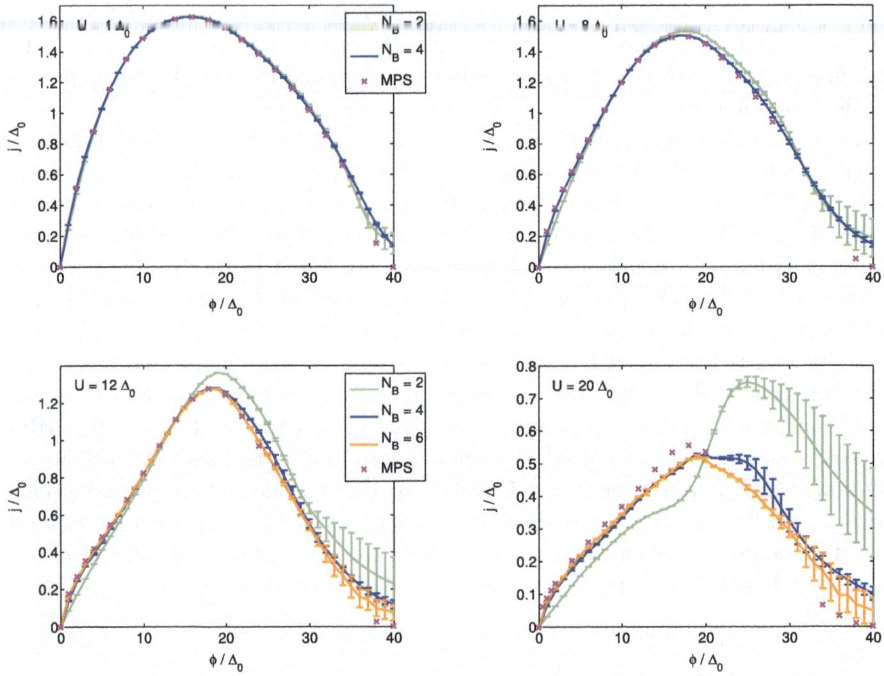

Fig. 4.16 ED results for the current-voltage characteristics of the nonequilibrium SIAM with semi-circular lead DOS. At the lead bandwidth $\phi = 40\,\Delta_0$, the current is strongly suppressed and must vanish for $U = 0$. MPS refers to quasi exact reference data [61]. Results taken from [63]

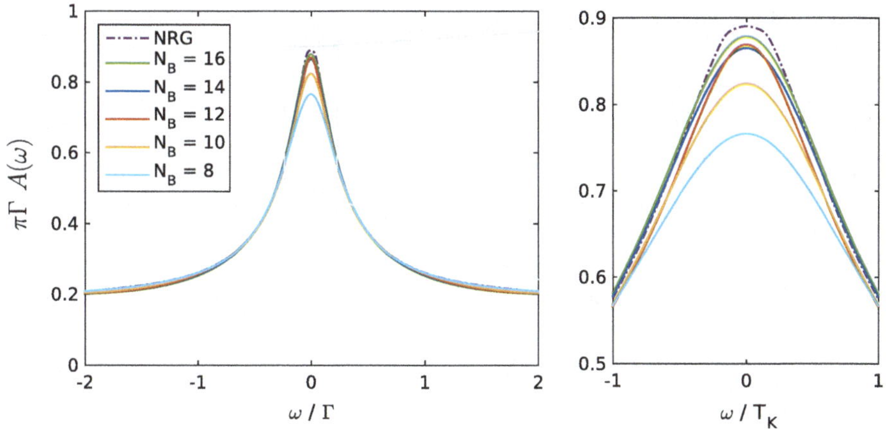

Fig. 4.17 MPS results for the equilibrium, i.e. $\phi = 0$, spectral function with increasing number of bath sites. A comparison to a quasi exact NRG reference calculation reveals a remarkably close agreement. Results for a temperature $T = 0.05\Gamma$ well below the Kondo scale $T_K \approx 0.2\Gamma$, taken from [78] (For $T/T_K \to 0$ the exact spectral function fullfills the so-called Friedel sum rule $A(\omega) = \pi\Gamma$.)

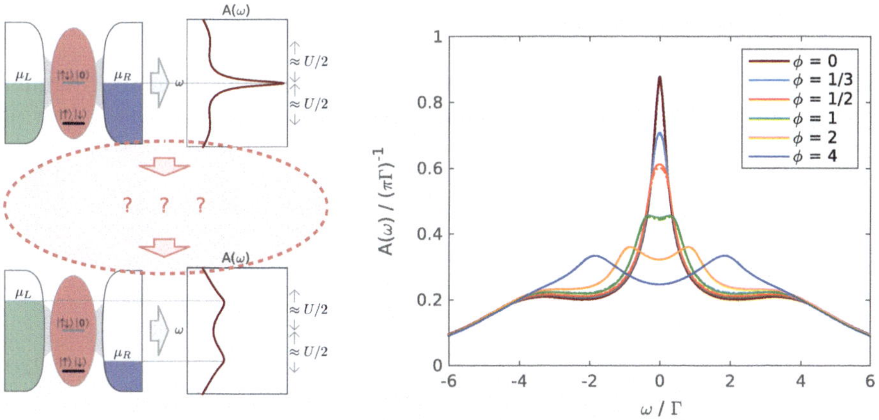

Fig. 4.18 Sketch of the nonequilibrium impurity problem on the left. In the low energy limit $\phi \ll T_K$, as well as in the high energy limit $\phi \gg T_K$, the detailed physics is known. On the right we present MPS results for the nonequilibrium spectral function in the challenging intermediate regime $T < T_K$ and $\phi \sim T_K$, displaying a clear splitting of the Kondo peak for $\phi > \Gamma/2$. Figure on the right taken from [78]

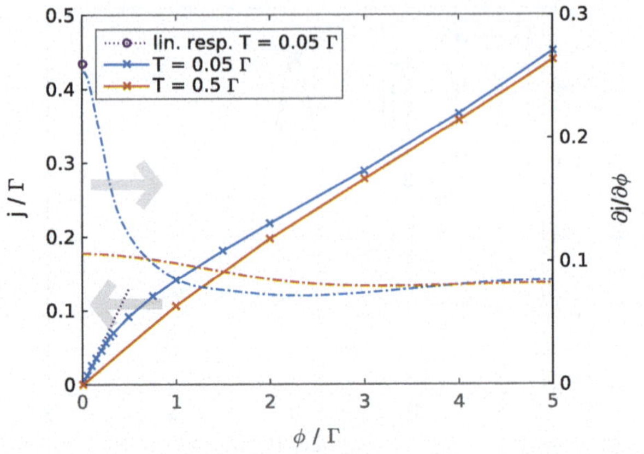

Fig. 4.19 MPS results for the current-voltage characteristics together with the differential conductance $\partial j/\partial \phi$. At low temperatures and close to $\phi \approx 0$, a clear enhancement of the conductivity due to the Kondo effect is found. Figure taken from [78]

$\underline{\Delta}^{Aux}(\omega)$, and thus, for fitting $\underline{\Delta}^{Aux}(\omega)$ to $\underline{\Delta}(\omega)$ by minimizing (4.218). After that we focus on the manybody problem and the computation of the interacting Green's function $\underline{G}^{Aux}$.

4.9.1.1 Noninteracting Case

We start from (4.153), (4.154) and (4.155), i.e. the Lindblad equation in superfermion representation. This equation of "Schrödinger type" can be rewritten in the following form

$$i\mathcal{L} = \underline{\underline{C}}^{\dagger}\underline{\underline{H}}\,\underline{\underline{C}} + \text{const.}, \tag{4.219}$$

where

$$\underline{\underline{C}} = \begin{pmatrix} c_1 \\ \vdots \\ c_N \\ \tilde{c}_1^{\dagger} \\ \vdots \\ \tilde{c}_N^{\dagger} \end{pmatrix} \tag{4.220}$$

summarizes the fermionic operators c_i ($\tilde{c}_i^{\dagger}$) for original (tilde) sites, and

$$\begin{aligned} \underline{\underline{H}} &= \begin{pmatrix} E_+ & B \\ \bar{B} & E_- \end{pmatrix}, \\ E_{\pm} &= E \pm i(\Gamma_2 - \Gamma_1), \\ B &= 2\Gamma_2, \\ \bar{B} &= -2\Gamma_1. \end{aligned} \tag{4.221}$$

The matrix E accounts hereby for all single-particle terms in the original Hamiltonian, i.e. all hoppings and onsite energies, and the dense matrices $\Gamma_{1/2}$ contain all the couplings to the Markovian environments, see (4.217).

The detailed derivation of the Green's functions for the generic case of a lattice model is rather lengthy and can be found for instance in [63, 95]. On the other hand, the final expression is quite compact, so we start by displaying it here in matrix form in the i, j indices

$$
\begin{aligned}
G_R(\omega) &= (\omega - E + i\Gamma_+)^{-1}, \\
G_K(\omega) &= 2i\, G_R(\omega)\Gamma_- G_A(\omega),
\end{aligned} \tag{4.222}
$$

with the abbreviations $\Gamma_\pm = \Gamma_2 \pm \Gamma_1$. These analytic expressions involve only matrix multiplications of size $N = N_B + 1$ and are thus numerically cheap to evaluate. We now prove (4.222).

Proof for the Retarded Component

The retarded Green's function is given by

$$
\begin{aligned}
G_R(t)_{\alpha\beta} &= -i\theta(t)\langle\{c_\alpha(t), c_\beta^\dagger\}\rangle \\
&= \theta(t)(p_{\alpha\beta}(t) + g_{\alpha\beta}(t)),
\end{aligned} \tag{4.223}
$$

with the first part

$$
\begin{aligned}
p_{\alpha\beta}(t) &= -i\mathrm{Tr} c_\alpha (c_\beta^\dagger \rho)_t = -i\langle I|c_\alpha (c_\beta^\dagger \rho)_t|I\rangle \\
&= -i\langle I|c_\alpha e^{\mathcal{L}t} c_\beta^\dagger|\rho\rangle \equiv -i\langle I|c_\alpha|c_\beta^\dagger(t)\rangle,
\end{aligned} \tag{4.224}
$$

where ρ is in the steady state, and the usual restriction $t > 0$ of the Lindblad formalism applies. When making use of the property $\langle I|\mathcal{L} = 0$ one can write the time derivative in terms of an equation of motion

$$
\frac{d}{dt} p_{\alpha\beta}(t) = -i\langle I|[c_\alpha, \mathcal{L}]|c_\beta^\dagger(t)\rangle. \tag{4.225}
$$

With the Lindblad operator in the form of (4.219) one finds for the commutator

$$
\begin{aligned}
[c_\alpha, i\mathcal{L}] &= \sum_{\tilde{\gamma}=1}^{2N} \underline{\underline{H}}_{\alpha\tilde{\gamma}}\, \underline{\underline{C}}_{\tilde{\gamma}} \\
&= \sum_{\gamma=1}^{N} \left(E_{+\alpha\gamma}\, c_\gamma + B_{\alpha\gamma}\, \tilde{c}_\gamma^\dagger\right),
\end{aligned} \tag{4.226}
$$

where we made use of the definition (4.220) in the second line. Multiplied with the left vacuum this results in terms of the form

$$\langle I|[c_\alpha, i\mathcal{L}] \to \sum_\gamma (E_+ - iB)_{\alpha\gamma} \langle I|c_\gamma = \sum_\gamma (E - i\Gamma_+)_{\alpha\gamma} \langle I|c_\gamma, \tag{4.227}$$

due to the tilde conjugation rule $\langle I|\tilde{c}^\dagger = -i\langle I|c$. In matrix notation one thus arrives at

$$\frac{d}{dt}p = -i(E - i\Gamma_+)p. \tag{4.228}$$

Now, we proceed analogously with the second part

$$g_{\alpha\beta}(t) = -i\mathrm{Tr}(c^\dagger_\beta c_\alpha(t)\rho) = -i\mathrm{Tr}c_\alpha(\rho c^\dagger_\beta)_t. \tag{4.229}$$

In terms of Lindblad time evolution and with the help of the tilde conjugation rules we find that

$$\begin{aligned} g_{\alpha\beta}(t) &= -i\langle I|c_\alpha e^{\mathcal{L}t}\rho c^\dagger_\beta|I\rangle = -i\langle I|c_\alpha e^{\mathcal{L}t}(-i\tilde{c}_\beta)|\rho\rangle \\ &= -\langle I|c_\alpha e^{\mathcal{L}t}\tilde{c}_\beta|\rho\rangle = -\langle I|c_\alpha|\tilde{c}_\beta(t)\rangle. \end{aligned} \tag{4.230}$$

In the same manner as before, by writing down the equations of motion one arrives at

$$\frac{d}{dt}g = -i(E - i\Gamma_+)g. \tag{4.231}$$

Finally, from inserting the results (4.228) and (4.231) into the time derivative of (4.223), one finds the following equation of motion for the retarded Green's function

$$i\frac{d}{dt}G_R(t) = I\delta(t) + (E - i\Gamma_+)G_R(t). \tag{4.232}$$

Fourier transforming of the lhs and rhs of the equation yields $\int i\frac{d}{dt}G_R(t)e^{i\omega t}dt = \omega G_R(\omega)$ and $I + (E - i\Gamma_+)G_R(\omega)$. Overall this results in the first equation in (4.222). The effective broadening Γ_+ corresponds in real time to a damping: $G_R(t) = \int_{-\infty}^{+\infty} G_R(\omega)e^{-i\omega t}\frac{d\omega}{2\pi} \sim e^{-\Gamma_+ t}$ for $t > 0$.

In analogous manner one obtains for the advanced Green's function the usual relation

$$\begin{aligned} G_A(\omega) &= G_R(\omega)^\dagger \\ &= (\omega - E - i\Gamma_+)^{-1}. \end{aligned} \tag{4.233}$$

Proof for the Keldysh component

For the Keldysh component we follow closely the one presented in [95]. Analogous to (4.223), G_K is given by

$$G_K(t) = -i\langle [c(t), c^\dagger] \rangle \\ = p(t) - g(t), \tag{4.234}$$

when written in matrix form. Using the expressions for $p(t)$ and $g(t)$ derived above, one has for $t > 0$

$$i\frac{d}{dt} G_K(t) = (E - i\Gamma_+) G_K(t), \tag{4.235}$$

with the solution

$$G_K(t > 0) = e^{-i(E - i\Gamma_+)t} \underbrace{G_{K0}}_{G_K(t=0)} . \tag{4.236}$$

For $t < 0$ one can use the following property

$$G_K(t) = -G_K(-t)^\dagger, \tag{4.237}$$

which is easily verified by inserting the definition of G_K, (4.234). For negative times one thus has

$$G_K(t < 0) = G_{K0} e^{-i(E + i\Gamma_+)t}, \tag{4.238}$$

since $G_{K0}^\dagger = -G_{K0}$. When splitting the time integration in the Fourier transform, one finds from (4.236) and (4.238)[22]

$$G_K(\omega) = \int_{-\infty}^{+\infty} dt e^{i\omega t} G_K(t) \\ = i(\omega - E + i\Gamma_+)^{-1} G_{K0} - iG_{K0}(\omega - E - i\Gamma_+)^{-1} \\ = i(G_R(\omega)G_{K0} - G_{K0}G_A(\omega)). \tag{4.239}$$

From the Langreth rules, (4.190), one has for the inverse

$$[\underline{G}(\omega)^{-1}]_K = -i(G_{K0}G_A(\omega)^{-1} - G_R(\omega)^{-1}G_{K0}), \tag{4.240}$$

and G_{K0} is given by the equal time expectation value

$$G_{K0} = -i(2m - I), \\ m_{\alpha\beta}(t) = \langle c_\alpha c_\beta^\dagger \rangle_t. \tag{4.241}$$

From the corresponding equation of motion we now determine the steady state expression for $m_{\alpha\beta}(t)$:

[22] Cf. Generalized Kadanoff-Baym ansatz (GKBA) [99].

$$0 = \frac{dm_{\alpha\beta}}{dt} = \frac{d}{dt}\langle c_\alpha c_\beta^\dagger \rangle = \langle I|[c_\alpha c_\beta^\dagger, \mathcal{L}]|\rho(t)\rangle = \langle I|(c_\alpha[c_\beta^\dagger, \mathcal{L}] + [c_\alpha, \mathcal{L}]c_\beta^\dagger)|\rho\rangle. \tag{4.242}$$

For the sake of clarity we omit matrix indices. Using (4.219) for the Lindblad operator we obtain

$$\begin{aligned}\frac{dm}{dt} &= -i\langle I|c(-\underline{\underline{C}}^\dagger \underline{\underline{H}})|\rho\rangle - i\langle I|(\underline{\underline{H}}\,\underline{\underline{C}})c^\dagger|\rho\rangle \\ &= i\langle I|c(c^\dagger E_+ + \tilde{c}\bar{B})|\rho\rangle - i\langle I|(E_+ c + B\tilde{c}^\dagger)c^\dagger|\rho\rangle \\ &= imE_+ - \underbrace{\langle I|c^\dagger c|\rho\rangle}_{1-m}\,\bar{B} - iE_+ m - Bm,\end{aligned} \tag{4.243}$$

where in the last line we have made use of the tilde conjugation rules. On the whole, the steady-state single particle density matrix $1 - m$ is obtained from solving

$$(m-1)\bar{B} - Bm - i[E_+, m] = 0. \tag{4.244}$$

With the definitions from (4.221) and (4.241) this amounts to

$$2\Gamma_- = (E - i\Gamma_+)G_{K0} - G_{K0}(E + i\Gamma_+). \tag{4.245}$$

The rhs is just what is obtained from (4.240), when inserting the expressions for $G_{R/A}(\omega)$:

$$\begin{aligned}[\underline{G}(\omega)^{-1}]_K &= -i(G_{K0}G_A(\omega)^{-1} - G_R(\omega)^{-1}G_{K0}) \\ &= -i(G_{K0}(\omega - E - i\Gamma_+) - (\omega - E + i\Gamma_+)G_{K0}) \\ &= 2i\Gamma_-.\end{aligned} \tag{4.246}$$

4.9.1.2 Interacting Case

To solve the interacting problem we set up the full manybody basis in the superfermion representation, which corresponds to the manybody Hilbert space for (density) operators of the $N = N_B + 1$ sites system. The Hilbert space size is exponentially large, $N_{\mathcal{H}} = 16^N$, when taking both spin directions and no particle conservation into account. In the following we assume that the Lindblad operator and the corresponding manybody states are expressed in this basis, so that the time evolution equation

$$|\rho(t)\rangle = e^{\hat{\mathcal{L}}t}|\rho(0)\rangle, \tag{4.247}$$

corresponds to a linear algebra problem of size $N_{\mathcal{H}}$. As before, $i\hat{\mathcal{L}}$ plays the role of a non-Hermitian Hamiltonian.

We assume that $\hat{\mathcal{L}}$ can be diagonalized, a property which is not trivial for a non-Hermitian matrix but can be argued from a physical point of view. The left- and right-sided eigenvectors

$$\begin{aligned}\hat{\mathcal{L}}|\alpha R\rangle &= \mathcal{L}_\alpha|\alpha R\rangle,\\ \langle\alpha L|\hat{\mathcal{L}} &= \mathcal{L}_\alpha\langle\alpha L|\,,\end{aligned} \tag{4.248}$$

with eigenvalues $\mathcal{L}_\alpha$ can be chosen in such a way that they are bi-orthogonal to each other

$$\langle\alpha L|\beta R\rangle = \delta_{\alpha\beta}, \tag{4.249}$$

and form a complete set,

$$\hat{I} = \sum_\alpha |\alpha R\rangle\langle\alpha L|. \tag{4.250}$$

Due to this we can expand any state $|\rho(t)\rangle$ in this basis and one can rewrite (4.247) as

$$\begin{aligned}|\rho(t)\rangle &= \sum_\alpha P_\alpha(t)|\alpha R\rangle,\\ P_\alpha(t) &= e^{\mathcal{L}_\alpha t}P_\alpha(0).\end{aligned} \tag{4.251}$$

For a stable solution one must have that $\Re\{\mathcal{L}_\alpha\} \leq 0$, which is ensured by the form of the Lindblad equation. Furthermore, at least one eigenvalue must be zero due to the property $\langle I|\hat{\mathcal{L}} = 0$ of the left vacuum, (4.157). We assume here that exactly one eigenvalue, say $\mathcal{L}_{\alpha=0}$, is zero with the following two corresponding left and right eigenstates

$$\begin{aligned}\langle I| &= \langle\alpha = 0, L|,\\ |\rho_\infty\rangle &= |\alpha = 0, R\rangle,\end{aligned} \tag{4.252}$$

where $|\rho_\infty\rangle$ is the steady state. From (4.249) it follows that

$$\langle I|\rho_\infty\rangle = 1, \tag{4.253}$$

which is just the trace normalization of the density matrix. The assumption that only one eigenvalue is zero, and, more generally, that only one has a vanishing real part, is expected to be fullfilled for systems in which each level which is connected to a Lindblad coupling. In this case, any initial state is expected to dissipate so that a unique steady state is fullfilled.

In order to reduce the relevant Hilbert space size $N_{\mathcal{H}}$ it is expedient to make use of conserved quantities. In many interesting cases, the number of particles per spin component is conserved. Within the superfermion representation this translates into the conservation of the operator[23]

$$\begin{aligned} \Delta_\sigma &= N_\sigma - \tilde{N}_\sigma \\ &= \sum_i \left(c^\dagger_{i\sigma} c_{i\sigma} - \tilde{c}^\dagger_{i\sigma} \tilde{c}_{i\sigma} \right) \end{aligned} \tag{4.254}$$

The left vacuum $\langle I|$ and, consequently also the steady state $|\rho_\infty\rangle$ are situated in the sector $\Delta_\sigma = 0$, as can be easily checked. The excited states used to evaluate Green's function belong to sectors with one of the $\Delta_\sigma \neq 0$. For example, $|c^\dagger_\uparrow(t)\rangle = e^{\mathcal{L}t} c^\dagger_\uparrow |\rho_\infty\rangle$ has $\Delta_\uparrow = 1$ and $\Delta_\downarrow = 0$. A general non-stationary state $|\rho(t)\rangle$ which is not an eigenstate of $\mathcal{L}$, however, does not necessarily have a well-defined particle sector Δ_σ. From (4.251) it follows that in this case all components with $\Delta_\sigma \neq 0$ are exponentially damped and only the steady state component $|\alpha = 0, R\rangle$ in the sector $\Delta_\sigma = 0$ survives in the long-time limit. Thus, for the purpose of finding the steady state $|\rho_\infty\rangle$, a convenient choice for the initial state is for instance $|\rho(t = 0)\rangle = |I\rangle$.[24]

Steady State Correlation Functions

Here we show that correlation functions in the steady state can be expressed in the form of a Lehmann representation, analogous to the equilibrium case (cf. [63, 100]). The steady state Green's function in the time domain, $iG_{BA}(t) \equiv \langle B(t)A\rangle$, reads in terms of the superfermion representation

$$\begin{aligned} iG_{BA}(t, +) &\equiv \theta(t)\langle I|B|A(t)\rangle \\ &= \theta(t)\langle I|Be^{\mathcal{L}t}A|\rho_\infty\rangle, \end{aligned} \tag{4.255}$$

where $+$ indicates that the time argument is ≥ 0 and one can use the "normal" quantum regression theorem, see Sect. 4.6. When inserting the identity operator in terms of the eigenstates of $\mathcal{L}$, (4.250), one obtains

$$\begin{aligned} iG_{BA}(t, +) &= \theta(t)\langle 0L|BIe^{\mathcal{L}t}IA|0R\rangle \\ &= \theta(t) \sum_\alpha \langle 0L|B|\alpha R\rangle \langle \alpha L|A|0R\rangle e^{\mathcal{L}_\alpha t}\,, \end{aligned} \tag{4.256}$$

[23]Also the analogue to the usual SU(2) spin symmetry can be implemented, although it is more involved.

[24]Note that a more standard form for the conserved quantities is obtained when performing a particle-hole transformation in the tilde space $\tilde{c}_\sigma \to \tilde{h}^\dagger_\sigma$, since then $\Delta_\sigma \to N_\sigma + \tilde{N}_\sigma - N$.

and subsequent Fourier transformation yields

$$
\begin{aligned}
G_{BA}(\omega,+) &= \int_{-\infty}^{\infty} G_{BA}(t,+)e^{i\omega t}dt \\
&= \sum_{\alpha} \langle 0L|B|\alpha R\rangle \langle \alpha L|A|0R\rangle \frac{1}{\omega - i\mathcal{L}_\alpha}.
\end{aligned} \tag{4.257}
$$

In particular we are interested in the retarded and Keldysh Green's functions. The retarded Green's function in the time domain is given by

$$
\begin{aligned}
G_{RBA}(t) &= -i\theta(t)\langle\{B(t), A\}\rangle \\
&= -i\theta(t)\langle B(t)A + A(-t)B\rangle \\
&= G_{BA}(t,+) + G_{AB}(-t,-),
\end{aligned} \tag{4.258}
$$

where the QRT cannot be applied directly to the second term. But, as before, from the complex conjugate one finds

$$
\begin{aligned}
iG_{AB}(-t,-) &\equiv \theta(t)\langle AB(t)\rangle \\
&= \theta(t)\langle B^\dagger(t)A^\dagger\rangle^* \\
&= \theta(t)\langle I|B^\dagger|A^\dagger(t)\rangle^* \\
&= \theta(t)\left(\sum_{\alpha} \langle 0L|B^\dagger|\alpha R\rangle \langle \alpha L|A^\dagger|0R\rangle e^{\mathcal{L}_\alpha t}\right)^*,
\end{aligned} \tag{4.259}
$$

and the Fourier transform is given by

$$
\begin{aligned}
G_{AB}(\omega,-) &= \int_{-\infty}^{\infty} G_{AB}(-t,-)e^{i\omega t}dt \\
&= \sum_{\alpha} \left(\langle 0L|B^\dagger|\alpha R\rangle \langle \alpha L|A^\dagger|0R\rangle\right)^* \frac{1}{\omega - i\mathcal{L}^*_\alpha}.
\end{aligned} \tag{4.260}
$$

On the whole, $G_{RBA}(\omega)$ is obtained by the Fourier transform of (4.258), so by the sum of (4.257) and (4.260). In particular, one can see that the poles are located at $\omega = i\mathcal{L}_\alpha$ and $\omega = i\mathcal{L}^*_\alpha$, so in the lower complex half plane since $\Re\{\mathcal{L}_\alpha\} \leq 0$ for all eigenstates. This ensures the causality of the retarded Green's functions. Due to these poles away from the real axis, the spectrum is not given by a sum of delta peaks, as for a finite size system, but by a continuous function.

The Lehmann representation of $G_{KAB}(\omega)$ is obtained in a similar way, leading to

$$\begin{aligned} G_{KAB}(\omega) &= -i\mathcal{F}\{\langle[B(t), A]\rangle\} \\ &= \sum_\alpha \langle 0L|B|\alpha R\rangle\langle\alpha L|A|0R\rangle \frac{1}{\omega - i\mathcal{L}_\alpha} \\ &+ \sum_\alpha \langle 0L|A|\alpha R\rangle\langle\alpha L|B|0R\rangle \frac{1}{\omega + i\mathcal{L}_\alpha} - \text{h.c.} \end{aligned} \tag{4.261}$$

with $\mathcal{F}\{\}$ denoting the Fourier transformation and h.c. stands for the Hermitian conjugate of the two sums.

In practice, the Green's functions described above can be determined either directly by full diagonalization of the Lindblad operator, see (4.248), or with Krylov space methods. A full diagonalization is only feasible for rather small matrices of size $N_\mathcal{H} \lesssim 5000$, due to memory constraints and since the numerical effort scales with the third power of the matrix size. Krylov space methods, on the other hand, allow one to consider much larger values of $N_\mathcal{H} \approx 10^7 - 10^9$, because only matrix-vector products are needed. Instead of computing all eigenvectors of the Lindblad operator, only the relevant subset, the so-called Krylov subspace is targeted in an iterative fashion. For the case of Green's functions, for instance, $\langle l|1/(\omega - \alpha\mathcal{L})|r\rangle$ is computed by forming a bi-orthogonal set of vectors spanned by $\langle l|\mathcal{L}^n$ and $\mathcal{L}^n|r\rangle$ with $n = 0, 1, 2, \ldots$. For more details on the common Lanczos algorithm for Hermitian problems, as well as the bi- or two-sided Lanczos and the Arnoldi algorithm for non-Hermitian matrices, which must be used in the present problem, we refer to [101–103].

Acknowledgements We gratefully acknowledge fruitful discussions with Michael Knap, Delia Fugger, Max Sorantin, Irakli Titvinidze, Frauke Schwarz, Jan von Delft, Sebastian Diehl, Wolfgang von der Linden, Hans Gerd Evertz, and Martin Nuss. Special thanks to Manuel Alamo, Fabio Covito, Daniel May, Matthias Peschke, and Christian Schäfer for help with the preparation of the manuscript. Thanks also to Roberta Citro and Ferdinando Mancini for the organization of this school. This work was partially supported by the Austrian Science Fund (FWF) within Projects P26508, and F41 (SFB ViCoM), as well as NaWi Graz. The calculations were partly performed on the D-Cluster Graz and on the VSC-3 cluster Vienna.

References

1. H. Haug, A.P. Jauho, *Quantum Kinetics in Transport and Optics of Semiconductors* (Springer, Heidelberg, 1998). http://www.springer.com/us/book/9783540735618
2. J. Rammer, H. Smith, Rev. Mod. Phys. **58**, 323 (1986). https://doi.org/10.1103/RevModPhys.58.323
3. P.W. Anderson, Phys. Rev. **124**, 41 (1961). https://doi.org/10.1103/PhysRev.124.41
4. J. Kondo, Progress Theor. Phys. **32**(1), 37 (1964). https://doi.org/10.1143/PTP.32.37. http://ptp.ipap.jp/link?PTP/32/37/
5. J.R. Schrieffer, P.A. Wolff, Phys. Rev. **149**, 491 (1966). https://doi.org/10.1103/PhysRev.149.491

6. A.C. Hewson, *The Kondo Problem to Heavy Fermions* (Cambridge University Press, Cambridge, 1993). https://doi.org/10.1017/CBO9780511470752 (Cambridge Books Online)
7. R. Bulla, T.A. Costi, T. Pruschke, Rev. Mod. Phys. **80**(2), 395 (2008). https://doi.org/10.1103/RevModPhys.80.395
8. M. Caffarel, W. Krauth, Phys. Rev. Lett. **72**(10), 1545 (1994). https://doi.org/10.1103/PhysRevLett.72.1545
9. A. Georges, G. Kotliar, W. Krauth, M.J. Rozenberg, Rev. Mod. Phys. **68**, 13 (1996). https://doi.org/10.1103/RevModPhys.68.13
10. H. Aoki, N. Tsuji, M. Eckstein, M. Kollar, T. Oka, P. Werner, Rev. Mod. Phys. **86**, 779 (2014). https://doi.org/10.1103/RevModPhys.86.779
11. P. Schmidt, H. Monien, Nonequilibrium dynamical mean-field theory of a strongly correlated system (2002). http://arxiv.org/abs/cond-mat/0202046
12. J.K. Freericks, V.M. Turkowski, V. Zlatić, Phys. Rev. Lett. **97**(26), 266408 (2006). https://doi.org/10.1103/PhysRevLett.97.266408
13. J.K. Freericks, Phys. Rev. B **77**(7), 075109 (2008). https://doi.org/10.1103/PhysRevB.77.075109
14. A.V. Joura, J.K. Freericks, T. Pruschke, Phys. Rev. Lett. **101**(19), 196401 (2008). https://doi.org/10.1103/PhysRevLett.101.196401. http://link.aps.org/abstract/PRL/v101/e196401
15. M. Moeckel, S. Kehrein, Phys. Rev. Lett. **100**, 175702 (2008). https://doi.org/10.1103/PhysRevLett.100.175702
16. M. Eckstein, M. Kollar, P. Werner, Phys. Rev. Lett. **103**, 056403 (2009). https://doi.org/10.1103/PhysRevLett.103.056403
17. S. Okamoto, Phys. Rev. B **76**, 035105 (2007). https://doi.org/10.1103/PhysRevB.76.035105
18. S. Okamoto, Phys. Rev. Lett. **101**, 116807 (2008). https://doi.org/10.1103/PhysRevLett.101.116807
19. A. Amaricci, C. Weber, M. Capone, G. Kotliar, Phys. Rev. B **86**, 085110 (2012). https://doi.org/10.1103/PhysRevB.86.085110
20. E. Munoz, C.J. Bolech, S. Kirchner, Phys. Rev. Lett. **110**, 016601 (2013)
21. E. Arrigoni, M. Knap, W. von der Linden, Phys. Rev. Lett. **110**, 086403 (2013). https://doi.org/10.1103/PhysRevLett.110.086403
22. I. Titvinidze, A. Dorda, W. von der Linden, E. Arrigoni, Phys. Rev. B **92**, 245125 (2015). https://doi.org/10.1103/PhysRevB.92.245125
23. A. Dorda, I. Titvinidze, E. Arrigoni, J. Phys.: Conf. Ser. **696**(1), 012003 (2016). https://doi.org/10.1088/1742-6596/696/1/012003
24. G. Mazza, A. Amaricci, M. Capone, M. Fabrizio, Phys. Rev. Lett. **117**, 176401 (2016). https://doi.org/10.1103/PhysRevLett.117.176401
25. M. Rigol, Phys. Rev. A **80**, 053607 (2009). https://doi.org/10.1103/PhysRevA.80.053607
26. J.H. Shirley, Phys. Rev. **138**, B979 (1965). https://doi.org/10.1103/PhysRev.138.B979
27. R. Citro, N. Andrei, Q. Niu, Phys. Rev. B **68**, 165312 (2003). https://doi.org/10.1103/PhysRevB.68.165312
28. A. Russomanno, S. Pugnetti, V. Brosco, R. Fazio, Phys. Rev. B **83**, 214508 (2011). https://doi.org/10.1103/PhysRevB.83.214508
29. A.K. Eissing, V. Meden, D.M. Kennes, Phys. Rev. Lett. **116**, 026801 (2016). https://doi.org/10.1103/PhysRevLett.116.026801
30. N.S. Wingreen, K.W. Jacobsen, J.W. Wilkins, Phys. Rev. Lett. **61**(12), 1396 (1988). https://doi.org/10.1103/PhysRevLett.61.1396
31. J.X. Zhu, A.V. Balatsky, Phys. Rev. B **67**, 165326 (2003). https://doi.org/10.1103/PhysRevB.67.165326
32. T. Brandes, Phys. Rep. **408**(5–6), 315 (2005). https://doi.org/10.1016/j.physrep.2004.12.002. http://www.sciencedirect.com/science/article/B6TVP-4FCSDMG-2/2/f74e72e3efc0ee8b22029719949f565f
33. T. Yamamoto, K. Watanabe, Phys. Rev. Lett. **96**, 255503 (2006). https://doi.org/10.1103/PhysRevLett.96.255503

34. J. Loos, T. Koch, A. Alvermann, A.R. Bishop, H. Fehske, J. Phys. Condens. Matter **21**(39), 395601 (2009). http://stacks.iop.org/0953-8984/21/i=39/a=395601
35. M. Galperin, A. Nitzan, M.A. Ratner, Phys. Rev. B **75**, 155312 (2007). https://doi.org/10.1103/PhysRevB.75.155312
36. T.A. Costi, A.C. Hewson, V. Zlatic, J. Phys. Condens. Matter **6**(13), 2519 (1994). http://stacks.iop.org/0953-8984/6/i=13/a=013
37. K.G. Wilson, Rev. Mod. Phys. **47**, 773 (1975)
38. E. Gull, A.J. Millis, A.I. Lichtenstein, A.N. Rubtsov, M. Troyer, P. Werner, Rev. Mod. Phys. **83**, 349 (2011)
39. N.S. Wingreen, Y. Meir, Phys. Rev. B **49**, 11040 (1994). https://doi.org/10.1103/PhysRevB.49.11040
40. E. Lebanon, A. Schiller, Phys. Rev. B **65**, 035308 (2001). https://doi.org/10.1103/PhysRevB.65.035308
41. A. Rosch, J. Paaske, J. Kroha, P. Wölfle, Phys. Rev. Lett. **90**(7), 076804 (2003). https://doi.org/10.1103/PhysRevLett.90.076804
42. J. Paaske, A. Rosch, J. Kroha, P. Wölfle, Phys. Rev. B **70**(15), 155301 (2004). https://doi.org/10.1103/PhysRevB.70.155301
43. A. Rosch, J. Paaske, J. Kroha, P. Wölfle, J. Phys. Soc. Jpn. **74**(1), 118 (2005). https://doi.org/10.1143/JPSJ.74.118. http://jpsj.ipap.jp/link?JPSJ/74/118/
44. J.E. Han, R.J. Heary, Phys. Rev. Lett. **99**, 236808 (2007). https://doi.org/10.1103/PhysRevLett.99.236808
45. F.B. Anders, Phys. Rev. Lett. **101**(6), 066804 (2008). https://doi.org/10.1103/PhysRevLett.101.066804
46. F. Heidrich-Meisner, A.E. Feiguin, E. Dagotto, Phys. Rev. B **79**, 235336 (2009)
47. J. Eckel, F. Heidrich-Meisner, S.G. Jakobs, M. Thorwart, M. Pletyukhov, R. Egger, New. J. Phys. **12**, 043042 (2010)
48. L. Mayrhofer, M. Grifoni, Eur. Phys. J. B **56**(2), 107 (2007). https://doi.org/10.1140/epjb/e2007-00097-3
49. D. Darau, G. Begemann, A. Donarini, M. Grifoni, Phys. Rev. B **79**, 235404 (2009). https://doi.org/10.1103/PhysRevB.79.235404
50. R. Gezzi, T. Pruschke, V. Meden, Phys. Rev. B **75**, 045324 (2007). https://doi.org/10.1103/PhysRevB.75.045324
51. S.G. Jakobs, V. Meden, H. Schoeller, Phys. Rev. Lett. **99**(15), 150603 (2007). https://doi.org/10.1103/PhysRevLett.99.150603
52. H. Schoeller, Eur. Phys. J. Spec. Top. **168**, 179 (2009). https://doi.org/10.1140/epjst/e2009-00962-3
53. P. Werner, T. Oka, M. Eckstein, A.J. Millis, Phys. Rev. B **81**(3), 035108 (2010). https://doi.org/10.1103/PhysRevB.81.035108
54. M.A. Cazalilla, R. Citro, T. Giamarchi, E. Orignac, M. Rigol, Rev. Mod. Phys. **83**, 1405 (2011). https://doi.org/10.1103/RevModPhys.83.1405
55. M. Pletyukhov, H. Schoeller, Phys. Rev. Lett. **108**, 260601 (2012). https://doi.org/10.1103/PhysRevLett.108.260601
56. F. Reininghaus, M. Pletyukhov, H. Schoeller, Phys. Rev. B **90**, 085121 (2014). https://doi.org/10.1103/PhysRevB.90.085121
57. G. Cohen, E. Gull, D.R. Reichman, A.J. Millis, Phys. Rev. Lett. **112**, 146802 (2014). https://doi.org/10.1103/PhysRevLett.112.146802
58. M. Nuss, C. Heil, M. Ganahl, M. Knap, H.G. Evertz, E. Arrigoni, W. von der Linden, Phys. Rev. B **86**, 245119 (2012). https://doi.org/10.1103/PhysRevB.86.245119
59. D. Sexty, T. Gasenzer, J. Pawlowski, Phys. Rev. B **83**, 165315 (2011). https://doi.org/10.1103/PhysRevB.83.165315
60. R. Hützen, S. Weiss, M. Thorwart, R. Egger, Phys. Rev. B **85**, 121408 (2012). https://doi.org/10.1103/PhysRevB.85.121408
61. M. Nuss, M. Ganahl, H.G. Evertz, E. Arrigoni, W. von der Linden, Phys. Rev. B **88**, 045132 (2013). https://doi.org/10.1103/PhysRevB.88.045132

62. M. Knap, E. Arrigoni, W. von der Linden, Phys. Rev. B **88**, 054301 (2013). https://doi.org/10.1103/PhysRevB.88.054301
63. A. Dorda, M. Nuss, W. von der Linden, E. Arrigoni, Phys. Rev. B **89**, 165105 (2014). https://doi.org/10.1103/PhysRevB.89.165105
64. S. Bock, A. Liluashvili, T. Gasenzer, Phys. Rev. B **94**, 045108 (2016). https://doi.org/10.1103/PhysRevB.94.045108
65. A.A. Dzhioev, D.S. Kosov, J. Chem. Phys. **134**, 044121 (2011)
66. T. Prosen, New J. Phys. **10**, 043026 (2008)
67. T. Prosen, J. Stat. Mech. Theory Exp. **2010**(07), P07020 (2010). http://stacks.iop.org/1742-5468/2010/i=07/a=P07020
68. S.A. Gurvitz, Y.S. Prager, Phys. Rev. B **53**, 15932 (1996). https://doi.org/10.1103/PhysRevB.53.15932
69. A.A. Dzhioev, D.S. Kosov, J. Chem. Phys. **135**(17), 174111 (2011). https://doi.org/10.1063/1.3658736. http://link.aip.org/link/?JCP/135/174111/1
70. S. Ajisaka, F. Barra, C. Mejía-Monasterio, T. Prosen, Phys. Rev. B **86**, 125111 (2012). https://doi.org/10.1103/PhysRevB.86.125111
71. A. Dorda, M. Sorantin, W. von der Linden, E. Arrigoni, New J. Phys. **19**(6), 063005 (2017). https://doi.org/10.1088/1367-2630/aa6ccc. http://stacks.iop.org/1367-2630/19/i=6/a=063005
72. H.P. Breuer, Phys. Rev. A **75**, 022103 (2007). https://doi.org/10.1103/PhysRevA.75.022103
73. H.P. Breuer, J. Gemmer, M. Michel, Phys. Rev. E **73**, 016139 (2006). https://doi.org/10.1103/PhysRevE.73.016139
74. W.M. Zhang, P.Y. Lo, H.N. Xiong, M.W.Y. Tu, F. Nori, Phys. Rev. Lett. **109**, 170402 (2012). https://doi.org/10.1103/PhysRevLett.109.170402
75. M.P. Woods, R. Groux, A.W. Chin, S.F. Huelga, M.B. Plenio, J. Math. Phys. **55**(3) (2014). https://doi.org/10.1063/1.4866769
76. I. de Vega, D. Alonso, Rev. Mod. Phys. **89**, 015001 (2017). https://doi.org/10.1103/RevModPhys.89.015001
77. L.M. Sieberer, M. Buchhold, S. Diehl, Rep. Prog. Phys. **79**(9), 096001 (2016). http://stacks.iop.org/0034-4885/79/i=9/a=096001
78. A. Dorda, M. Ganahl, H.G. Evertz, W. von der Linden, E. Arrigoni, Phys. Rev. B **92**, 125145 (2015). https://doi.org/10.1103/PhysRevB.92.125145
79. U. Schollwöck, Ann. Phys. **326**(1), 96 (2011). https://doi.org/10.1016/j.aop.2010.09.012. http://www.sciencedirect.com/science/article/pii/S0003491610001752
80. F. Verstraete, J.J. García-Ripoll, J.I. Cirac, Phys. Rev. Lett. **93**(20), 207204 (2004). https://doi.org/10.1103/PhysRevLett.93.207204. http://link.aps.org/abstract/PRL/v93/e207204
81. M. Zwolak, G. Vidal, Phys. Rev. Lett. **93**(20), 207205 (2004). https://doi.org/10.1103/PhysRevLett.93.207205. http://link.aps.org/abstract/PRL/v93/e207205
82. T. Prosen, M. Znidaric, J. Stat. Mech. **2009**(02), P02035 (2009). http://stacks.iop.org/1742-5468/2009/i=02/a=P02035
83. H.P. Breuer, F. Petruccione, *The Theory of Open Quantum Systems* (Oxford University Press, Oxford, England, 2009). https://global.oup.com/academic/product/the-theory-of-open-quantum-systems-9780199213900
84. G. Schaller, *Open Quantum Systems Far from Equilibrium*. Lecture Notes in Physics (Springer, Heidelberg, 2014). http://www.springer.com/us/book/9783319038766
85. H.J. Carmichael, *Statistical Methods in Quantum Optics: Master Equations and Fokker-Planck Equations, Texts and Monographs in Physics*, vol. 1 (Springer, Singapore, 2002)
86. V. May, O. Kühn, *Charge and Energy Transfer Dynamics in Molecular Systems* (Wiley-VCH, Wennheim, 2011)
87. C.W. Gardiner, P. Zoller, *Quantum Noise* (Springer, Berlin, 2000)
88. C. Cohen-Tannoudji, J. Dupont-Roc, G. Grynberg, Atom-Photon Interactions: Basic Processes and Applications (Wiley-VCH, Weinheim, 2004). https://doi.org/10.1002/9783527617197
89. J. Dalibard, Y. Castin, K. Mølmer, Phys. Rev. Lett. **68**, 580 (1992). https://doi.org/10.1103/PhysRevLett.68.580

90. H.P. Breuer, B. Kappler, F. Petruccione, Phys. Rev. A **56**, 2334 (1997). https://doi.org/10.1103/PhysRevA.56.2334
91. H. Breuer, B. Kappler, F. Petruccione, Eur. Phys. J. B **1**(1), 9 (1998). https://doi.org/10.1007/s100530050058
92. A.J. Daley, J.M. Taylor, S. Diehl, M. Baranov, P. Zoller, Phys. Rev. Lett. **102**, 040402 (2009). https://doi.org/10.1103/PhysRevLett.102.040402
93. Y. Nakamura, Y. Yamanaka, Ann. Phys. **331**, 51 (2013). https://doi.org/10.1016/j.aop.2012.12.005. http://www.sciencedirect.com/science/article/pii/S0003491612002205
94. M. Schmutz, Z. Phys. B **30**, 97 (1978). https://doi.org/10.1007/BF01323673
95. F. Schwarz, M. Goldstein, A. Dorda, E. Arrigoni, A. Weichselbaum, J. von Delft, Phys. Rev. B **94**, 155142 (2016). https://doi.org/10.1103/PhysRevB.94.155142
96. R. van Leeuwen, N.E. Dahlen, G. Stefanucci, C.O. Almbladh, U. von Barth, *Introduction to the Keldysh Formalism* (Springer, Berlin, 2006), *Lecture Notes in Physics*, vol. 706, Chap. 3, pp. 33–59
97. E.N. Economou, *Green's Functions in Quantum Physics* (Springer, Heidelberg, 2006)
98. Y. Meir, N.S. Wingreen, Phys. Rev. Lett. **68**(16), 2512 (1992). http://link.aps.org/abstract/PRL/v68/p2512
99. K. Balzer, S. Hermanns, M. Bonitz, J. Phys. Conf. Ser. **427**(1), 012006 (2013). http://stacks.iop.org/1742-6596/427/i=1/a=012006
100. A.L. Fetter, J.D. Walecka, *Quantum Theory of Many-Particle Systems* (McGraw-Hill, New York, 1971)
101. Y. Saad, *Numerical Methods for Large Eigenvalue Problems, Revised Edition* (Society for Industrial and Applied Mathematics, 2011)
102. P. Arbenz. Lecture notes on solving large scale eigenvalue problems (2012). http://people.inf.ethz.ch/arbenz/ewp/lnotes.html. Online; Accessed May 2016
103. Z. Bai, J. Demmel, J. Dongarra, A. Ruhe, H. van der Vorst, *Templates for the Solution of Algebraic Eigenvalue Problems: A Practical Guide (Software, Environments and Tools)* (Society for Industrial and Applied Mathematics, 1987)

Index

R. Citro and F. Mancini (eds.), *Out-of-Equilibrium Physics of Correlated Electron Systems*, Springer Series in Solid-State Sciences 191,
https://doi.org/10.1007/978-3-319-94956-7